Anonymous

Bulbs, Plants and Seeds for Autumn Planting

In the Year 1897

Anonymous

Bulbs, Plants and Seeds for Autumn Planting
In the Year 1897

ISBN/EAN: 9783337372514

Printed in Europe, USA, Canada, Australia, Japan

Cover: Foto ©berggeist007 / pixelio.de

More available books at **www.hansebooks.com**

Peter Henderson & Co.

35 & 37 CORTLANDT ST.,

NEW YORK

BULBS.

PLANTS AND SEEDS.

1897

FOR AUTUMN PLANTING.

PETER HENDERSON & CO.'S

AUTUMN 1897 CATALOGUE

BULBS, PLANTS AND SEEDS.

HOW BEST TO ORDER AND REMIT.

Remittances should be made either in the form of an **Express Money Order** (*which is the best and safest method of remitting*), or a **P. O. Money Order, Bank Draft or Registered Letter.**

Customers can also shop by **express**, as all express companies have purchasing departments in all their offices. If you wish to order goods from us to be sent by express, ask for a Purchasing Order Blank at any express office, and it will be supplied and forwarded without charge, other than usual rates for returning goods.

If goods are wanted C. O. D., 25 per cent. of the amount must accompany the order.

Orders from unknown correspondents, without remittance, should be accompanied by New York City references, to save delay.

WE DELIVER FREE

TO ANY POST OFFICE OR R. R. EXPRESS OFFICE

IN THE UNITED STATES,

AT PRICES QUOTED IN THIS CATALOGUE, ALL

BULBS, VEGETABLE AND FLOWER SEEDS,

EXCEPT WHERE NOTED.

Purchaser pays Transit Charges only on Bulbs and Roots in quantity where specially noted ; also on Vegetable Seeds by the pint, quart, peck, bushel and barrel, and on Farm Seeds, Tools, Fertilizers, Insecticides, etc.

IN REGARD TO THE SHIPMENT OF

PLANTS,

SEE PAGE 57.

WE MAKE NO CHARGE FOR PACKING AND SHIPPING.

We make no charge for packing, cases, baskets, packages or barrels, and pack as lightly as possible, consistent with safety, so as to reduce the cost of carriage when paid by purchaser. In fact, we assume extra expense in supplying light, strong cases, and, where possible, we pack in baskets covered with waterproof covers, which greatly reduces weight. Cloth bags only we charge for, and these at cost. **All grass seed bags,** excepting bags for Timothy and Clover, are furnished free. We make no charge for cartage or delivery of goods to any railroad station, steamship line or express office in New York City proper. The safe arrival is guaranteed of all goods sent by express to any part of the United States or Canada. We ship plants at all seasons of the year, even in the coldest weather.

☞ *If your order for Bulbs includes some that are not ready before November, please state if you want us to send all of the other goods at once, and the November Bulbs as soon as ready—or if we shall hold the whole order and make one shipment.*

FREE | **PURCHASERS MAY SELECT PREMIUMS** | **PREMIUMS**

ON ORDERS FROM THIS CATALOGUE FOR

Bulbs (at individual prices, not in collections.), Plants, Vegetable and Flower Seeds,

AS FOLLOWS :

On $1.00 order Select Books, Bulbs, Seeds or Plants to value $1.10	On $8.00 order Select Books, Bulbs, Seeds or Plants to value $9.20
" 2.00 " " " " " " " 2.25	" 9.00 " " " " " " " 10.35
" 3.00 " " " " " " " 3.45	" 10.00 " " " " " " " 11.50
" 4.00 " " " " " " " 4.60	" 12.00 " " " " " " " 14.00
" 5.00 " " " " " " " 5.75	" 15.00 " " " " " " " 17.50
" 6.00 " " " " " " " 6.90	" 20.00 " " " " " " " 23.50
" 7.00 " " " " " " " 8.05	" 25.00 " " " " " " " 29.50

On a $50.00 order Select Books, Bulbs, Seeds or Plants to value $60.00.

ABOVE PREMIUMS CANNOT BE ALLOWED ON

Grass, Clover and Farm Seeds, Fertilizers, Insecticides, Tools,

OR ANY OF THE ARTICLES

OFFERED ON PAGES 75 to 80.

PREMIUMS | **FREE**

ALFRED HENDERSON, President.
CHARLES HENDERSON, Vice-Pres't and Treas.

PETER HENDERSON & CO.,
35 & 37 CORTLANDT STREET, NEW YORK.

INDEX.

BULBS. Pages 2 to 51.

PLANTS. Pages 52 to 68.

6 bulbs of one variety sold at dozen rates, 25 at 100 rates, 250 at 1,000 rates.

BULBS · THEIR USES AND How to Grow Them

ULBOUS PLANTS are among the most showy and useful of our garden favorites, are easily managed and are sure to bloom. The outdoor display may be fairly said to commence with March, when the garden is growing bright with Snow-drops, Scillas, Chionodoxas, Crocuses, Daffodils, etc., and during April and well into May the flower-beds are brilliant and charming with a wealth of lovely Hyacinths, Tulips, Anemones, Narcissus, etc., which make up a display of floral beauty rarely equaled. As winter-blooming plants they hold an important place, as there is no period of the year during which flowers are so highly appreciated, and certainly no class of flowering plants affords more pleasure. It is no difficult matter, by early planting and forcing a few Van Thol Tulips, Roman and other Hyacinths, Paper-white Narcissus, etc., to have them in bloom by New Year's day, while a few successive plantings of these and other choice sorts will ensure a beautiful display throughout the dull winter months.

... CULTURE OF HARDY BULBS, for flowering during the Winter, in the Window or Greenhouse. ...

Potting.—The bulbs should be planted in pots as soon as received. The soil should be rich and well mixed with at least one-third of old well-rotted manure; fill the pot nearly full of soil, place the bulb in, then fill in with soil firmly to within half an inch of the top of the pot.

After Potting.—One of the most important things to observe is the proper placing of the pots containing the bulbs. To get the best results in flowering the pots must be filled with roots before the top starts to grow, and to do this they must be plunged in some cool place, too much warmth excites the top into growth before the roots sufficient roots to nourish it. The most satisfactory method is to plunge the pots in the earth an inch or more below the surface, right out in the garden. Select a sheltered position, high enough so that water from rains will not settle and remain stagnant around the pots. When the weather gets cold enough to freeze the ground place three or four inches of

leaves, straw or other refuse over the soil where the pots are plunged. Bulbs should be sufficiently rooted in about eight or ten weeks after potting to have the pots lifted and brought in the house for flowering, though some kinds require a longer period. But to be sure that the roots have developed properly place the hand over the top of the pot, turn it upside down and tap the pot slightly, when the ball of earth will slip out of the pot; if the roots are plainly showing all around the earth the bulb is well rooted; place back in the pot and remove to the house for flowering whenever desired. By taking in a few pots at intervals of two weeks or more, a succession of bloom may be had throughout the winter. They should then be grown on in the house at a temperature of from sixty to seventy degrees, and should then bloom in from six to eight weeks from the time they are brought in after potting. Do not allow the plants to suffer for want of water, and if some Henderson's Plant Fertilizer or Manure water is added once a week it will be beneficial.

... GARDEN CULTURE OF HARDY BULBS. ...

To secure really fine flowers outdoor planting should be done early in the fall, though generally speaking, from October to the middle of November is the most desirable time. Most bulbs succeed in any well drained, good garden soil, which, however, should be dug at least eighteen inches deep. Hardy bulbs throw out their roots during the fall and winter—they usually root deeply; therefore the bulbs should be planted from three to four inches below the surface, so as to be as free as possible from the upper crust of the soil, which leaves considerably, caused by alternate freezing and thawing, thus causing bulbs planted too near the surface to break from their roots. A little sand placed below and around the bulbs permits the water to drain off in heavy

soils. Beds should be in a sunny position, if possible, and protected during winter by a coating of rotted manure. The beds may be taken up and dried off as soon as the leaves acquire a yellow color; the beds will be vacant in time for the ordinary bedding plants. If it is found necessary to remove the bulbs immediately after flowering they should be carefully taken up, the leaves and roots damaged as little as possible, and "heeled in" in some slightly shaded place until the foliage is quite withered and the bulbs thoroughly ripened, when they may be taken up, cleaned, and stored in a cold dry shed or cellar until wanted for the next fall's planting.

Bulbs for Geometrical Beds and Ribbon Borders.

Most showy and satisfactory effects are produced by planting the various colored Tulips, Hyacinths, etc., that grow about the same height and flower at the same time, in lines, each of one color, or in masses or geometrical designs, care being taken to arrange the various colors so the contrasts will be harmonious. We subjoin a few of the sorts usually used:

HYACINTHS (single varieties are preferred, as they produce better spikes). — **Pinks :** Gertrude, Norma, Gigantea. **Reds :** Robert Steiger, Veronica. **Whites :** Baroness von Thuyll, Grand Vedette, Voltaire. **Light Blues :** Charles Dickens, Czar Peter, La Peyrouse. **Dark Blues :** Baron von Thuyll, Marie.

SINGLE TULIPS.—**Crimsons :** Artus, Belle Alliance, Crimson King, Pott-bakker Scarlet, Vermillion Brilliant. **Rose and Pinks :** Cottage Maid, Rosa Mundi Huykman. **Yellows :** Canary Bird, Chrysolora, Pottebakker Yellow, Yellow Prince. **Whites :** Queen Victoria, L'Immaculee, Pottebakker White. **Clarets :** Wouverman, Vander Neer. **Variegated :** Kaiser-Kroon, Golden Standard, Grand Duc de Russie.

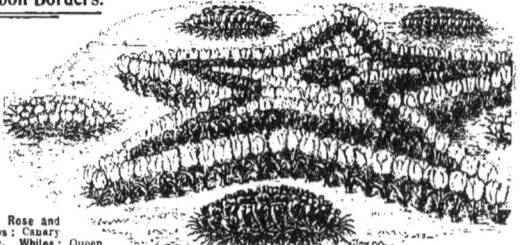

GEOMETRICAL BED OF EARLY TULIPS.

"NATURALIZING" HARDY BULBS
FOR PERMANENT EFFECTS IN
LAWNS AND GARDENS.

Beautiful and permanent effects may be obtained by planting hardy bulbs in groups and masses on the lawn, in shady nooks, where they find a congenial and permanent home, flowering abundantly in their season, and requiring little or no care after being planted. This mode of planting is termed "naturalizing," and is now generally followed in Europe. It adds a charm to tangled and half-wild places, heightens the natural effects of light and shade, and imparts a natural grace and beauty to the scene. The following are admirably adapted for this purpose:

Anemones, Apennina, Blanda and Hepatica. *Partial shade.*

Allium Moly. *Open and sunny position.*

Bulbocodium. *Partial shade.*

Camassia. *Shady woods.*

Chionodoxa. *Open or shady banks.*

Colchicum. *Open and sunny position.*

Crocus. *Open and sunny places.*

Eranthis (Winter Aconite). *Partial shade, under trees, etc.*

Erythronium. *Partial shade.*

Grape and Feather Hyacinths. *Partial.*

Hemerocallis. *Open, sunny, moist.*

Hyacinth. *Sheltered but open*

Iris Germanica. *Moist rich banks.*

Iris Kæmpferi. *Banks of streams, etc.*

Jonquils. *Open and sunny.*

Liliums. *Various sorts. Open and sunny position.*

Lily of the Valley. *Shady woods.*

Narcissus (Daffodil). *Open or shady.*

Pæonias. *Open and sunny.*

Puschkinia. *Partial shade.*

Scillas. *Shady banks and woods.*

Snowdrops. *Partial shade, under trees.*

Snowflakes. *Open or partial.*

Sternbergias. *Open and sunny.*

Triteleia. *Open and sunny.*

Trillium. *Shady woods.*

Tulips. *Open and sunny.*

Zephyranthes. *Open and sunny.*

BULBS FOR THE WINDOW GARDEN.

There is no class of plants that is more important or that gives more satisfaction for the window garden than those grown from bulbs. They are the most easily grown of all, and are sure to bloom abundantly. There is nothing more cheering or pleasant than a few Hyacinths, Tulips, Freesias, and other bulbs displaying their gorgeous colors and delightful fragrance during the dull months of winter. They should be potted as soon as received and placed to become thoroughly rooted, as described under "Forcing Bulbs." Interesting and beautiful objects for the window garden may be produced by growing Hyacinths in glass and by planting Crocuses or Lily of the Valley in the ornamental styles of pots made specially for the purpose, and pierced with holes to allow the shoots to escape and the flowers to expand. The following are of easy management, and will be found exceedingly attractive for the window garden during winter and spring:

Achimenes, Alliums (omitting the hardy sorts), **Amaryllis, Anemones** (omitting the hardy sorts), **Anomatheca, Arums, Babiana, Brodiæa, Bulbocodium, Callas, Calochortus, Chionodoxa, Crocus, Cyclamens, Freesia, Fritillaria, Gloxinias, Hyacinths, Iris** (Spanish), **Ixias, Jonquil, Lachenalia, Lily of the Valley, Liliums Candidum and Harrisii, Narcissus, Nerine, Ornithogalum, Oxalis, Schizostylis, Scilla Peruviana, Sparaxis, Triteleia, Tritonia, Tropæolum, Tulips (Early), Vallota.**

FORCING BULBS.

FOR WINTER FLOWERING IN THE CONSERVATORY OR GREENHOUSE.

There is no class of plants that gives the satisfaction and profusion of bloom in "winter forcing" that bulbs do, and with so little skill and care give such magnificent results. They occupy no space in the conservatory or window excepting when in the full wealth of bud and bloom, and then the effect of their incomparable flowers and rich gorgeous colors is matchless. As the proper way to treat them is to pot the bulbs in the autumn, or if for cutting only they may be planted an inch or two apart in shallow boxes about four inches deep; in either case plunge them in cold frames or in the open ground and cover with leaves, tan bark or something similar, and leave them until they become well rooted, which will be in about two months' time; they can then be brought in and the warmth and light of the conservatory or window will "force" them to bloom in two or three weeks' time. A continuous display of bloom may be enjoyed during the entire winter and spring by bringing in a few pots at intervals of two to three weeks. The following are among the best for forcing:

Alliums Neapolitanum and Grandiflorum, Anemone Fulgens, Callas, Freesias, Gladiolus Bride, Heleborus, Roman, Italian and Dutch Hyacinths, Scarlet Ixias, Campernelle Jonquils, Lily of the Valley Pips, Easter Liliums Harrisii, Candidum and Longiflorum, Narcissus, Ornithogalum, Astilbes, Spanish Iris, Single Early Tulips, Double Early Tulips.

.. HENDERSON'S ..
"RAINBOW" COLLECTION
OF BULBS.

For Winter Flowers in the Conservatory or Window Garden.

THIS collection is made up of the same choice bulbs that we offer in this Catalogue, and will give an endless amount of bloom throughout the winter. As they are offered at prices much below our Catalogue rates we consequently can make no alterations. Price, "Full Collection," containing 640 bulbs enumerated below, $10.00, or delivered in the United States, $10.75.

```
 20 Hyacinths, Splendid Named.
 48    "      Mixed, Single and Double.
 20    "      Roman, Assorted Kinds.
 24 Tulips, Named, Double and Single.
 48    "   Mixed Varieties.
 36 Narcissus, Splendid Named.
 48 Crocus, Splendid Named.
100    "    Mixed Sorts.
 20 Allium Neapolitanum.
 20 Anemone, Single and Double.
  4 Arum Dracunculus.
 20 Babianas, Mixed.
  4 Calla, White.
 20 Chionodoxa.
  4 Cyclamen Persicum.
 36 Freesia.
 20 Iris, Spanish, Mixed.
 20 Ixias, Mixed.
 36 Jonquils.
  4 Lily, Bermuda Easter.
  8 Ornithogalum Arabicum.
 20 Oxalis, Mixed.
 20 Ranunculus.
 20 Scilla Sibirica.
 20 Sparaxis, Mixed.
 20 Triteleia uniflora.
 20 Snowdrops.
```

"Half" of above collection of Winter-blooming Bulbs contains 340 bulbs. Price, $5.25, or delivered in the United States, $5.75.

"Quarter" of above collection of Winter-blooming Bulbs contains 170 bulbs. Price, $2.75, or delivered in the United States, $3.00.

.. HENDERSON'S ..
"RAINBOW" COLLECTION
OF HARDY BULBS.

For Spring Flowering in the Garden.

ALL are perfectly hardy and should be planted in the open ground this autumn, they will then give a grand and continuous display of bloom throughout the spring. As they are offered at much less than our regular Catalogue rates we can allow of no alterations. Price, "Full Collection," containing 900 bulbs, enumerated below, $10.00, or delivered in the United States, $10.75.

```
 60 Hyacinths, Mixed, Colors Separate.
 24    "       Grape and Feathered.
 60 Tulips, Single and Double, Named.
180    "   Mixed, Single and Double.
 20    "   Mixed Parrot.
 48 Narcissus, Splendid Mixed.
200 Crocus, Mixed, Colors Separate.
100    "   Named Sorts.
 12 Allium Moly.
 12 Bulbocodium vernum.
 48 Chionodoxa.
  4 Crown Imperial.
 48 Snowdrops.
  8 Fritillaria, Mixed.
 20 Iris, Assorted.
  8 Leucojum vernum.
 48 Scilla Amoena.
 12    "   Campanulata.
  8 Lilies, Splendid Sorts, all different.
```

"Half" of above collection of "Hardy" Bulbs, containing 450 bulbs, $5.25, or delivered in the United States, $5.75.

"Quarter" of above collection of "Hardy" Bulbs, containing 225 bulbs, $2.75, or delivered in the United States, $3.00.

"HENDERSON'S BULB CULTURE" gives full instructions for growing bulbs, given free with any of the above collections.

Novelties in Bulbs.

NEW AND RARE HYACINTHS

THE under-mentioned twelve varieties have been selected with great care from the newer Holland production in Hyacinths as being of superior merit. There are no two colors or shades alike in the set, and if grown under congenial conditions, either to flower in the house this winter or in the garden in spring, they are sure to give unbounded satisfaction.

SINGLE VARIETIES.

Bazine. Deep, rich crimson-scarlet, tall, well-furnished spike, the finest dark red.

Cardinal Wiseman. Dainty pink blush throughout, large, broad truss.

Captain Boynton. Light porcelain blue, very large bells, grand spike.

King of Yellows. Deep, pure yellow, grand truss.

Masterpiece. Magnificent dark purple-blue, compact truss.

Snowball. Very fine pure snowy white, extra large waxy bells.

DOUBLE VARIETIES.

Koh-i-noor. Deep, bright rose, large, semi-double bells, immense spike.

Minerva. Salmon, with rosy carmine stripe, fine double bells, compact rounded spike.

Robert Burns. Deep indigo black, double bells, compact spike.

Sir Joseph Paxton. Bright rosy carmine, large double bells, splendid truss.

Venus. Ivory white, with rosy blush centre, broad double bells, large spike.

Von Speyck. Deep lavender blue, with darker shadings; a splendid bright color, immense, very double broad bells, extra fine, tall, straight, compactly-furnished spike.

... Prices ...

Of New and Rare Hyacinths.

(Delivered Free in the United States.)

Any one variety, 25c. ; any 3, 65c. ; any 6 for $1.25 ; or the complete set of 6 doubles and 6 singles, all separately named, for $2.00.

NEW ROMAN HYACINTHS.

The flowers, while smaller than those of the Dutch Hyacinth, yet are produced in much greater abundance, each bulb producing several graceful spikes of bloom; their delicious perfume, earliness and profusion of bloom have made Roman Hyacinths exceedingly popular for winter flowering in the house.

Double White Romans.	10c. each,
Canary Yellow Romans.	85c. per doz .
	$5.00 per 100.

At the 100 price purchaser pays transit.

ROMAN HYACINTH.

Bulbs are delivered free in the United States, except where noted.   Avail yourself of the Premium offer on 2d page of cover.

NOVELTIES IN SINGLE EARLY FLOWERING TULIPS.

We have selected these Tulips with great care, from many of the newer sorts introduced by the Holland growers. They are all of exceptional beauty, and the sets comprise a wide range of color.

Gold Finch. *B6.* A large and showy tulip of rich bright yellow, retaining its bright color without turning brown. Very fragrant. 5c. each, 50c. per dozen.

Grand Duc d'Orange. *A8.* A surprisingly lovely flower for its decided scarlet flames on a rich yellow ground color. 5c. each, 50c. dozen.

John Bright. *A9.* Grand, extra large, broad petaled flower, bright rosy crimson, large yellow base, zoned white. 8c. each, 75c. per dozen.

La Matelas. *B9.* Immense grand flowers, 4 inches across, deep rose pink suffused lighter pink and blush, yellow base. 5c. each, 50c. per dozen.

La Remarquable. *B9.* Flower large, broad, and of fine form, color unique and rich, deep crimson lake, breaking into wide margin of blush pink. 5c. each, 50c. per dozen.

Maes. *B10.* One of the best tulips of recent introduction; a very large flower of intense dazzling scarlet. 5c. each, 50c. per dozen.

Moncheron. *B9.* Splendid large flower, rich dark scarlet, as if suffused with black blood, small yellow base, black anthers. 5c. each, 50c. per doz.

Ophir d'Or. *A9.* Grand large flower of deep rich yellow—the finest yellow grown. 8c. each, 75c. per dozen.

Rose Hawk. *A7.* A lovely delicate pink, large flower, extra fine for pots and forcing. 6c. each, 60c. per dozen.

White Eagle. *A8.* Large white, very slightly shaded with rose, robust habit, very early. 6c. each, 60c. per dozen.

Novelties in Early Flowering Double Tulips.

Blanche Hative. *A10.* A very fine and large semi-double, pure white, very early. 5c. each, 50c. per dozen.

Epaulette d'Argent. *B8.* A very striking flower. Red and white striped, large and double. 5c. each, 50c. per dozen.

Raphael. *A7.* The finest double Tulip grown, very large, blush white, shaded with darker rose. 8c. each, 75c. per dozen.

Parmesiano. *B8.* Extra large double flower of fine bright rose pink color. 8c. each, 75c. per dozen.

Toreador. *A7.* Orange red with rich yellow border. A splendid large very double flower. 8c. each, 75c. per dozen.

Vuurbaak. *A7.* Very brilliant vermilion scarlet, large. 5c. each, 50c. per dozen.

Collections OF NOVELTY TULIPS.

Delivered free in the U. S.

"Novelty" Double Tulips.

6 bulbs, 1 each of the above 6 varieties,			$0.35
18 "	3 "	" "	1.00
36 "	6 "	" "	1.85
72 "	12 "	" "	3.50

"Novelty" Single Tulips.

10 bulbs, 1 each of above 10 varieties,			$0.35
30 "	3 "	" "	1.50
60 "	6 "	" "	2.75
120 "	12 "	" "	5.00

BULBS ARE DELIVERED FREE IN THE U. S., EXCEPT WHERE NOTED. See our Premium Offer on Second page Cover.

DOUBLE TULIPS

GIANT GESNERIANA TULIPS.

GIANT GESNERIANA ...TULIPS

Belong to the late or May-flowering Garden Tulips. The flowers are very large, of symmetrical form, and are borne on tall, strong stems, often two feet high. They by far surpass in colors and brilliancy anything before known in Tulips. The colors are so glowing and bright that in the sunlight the effect is fairly dazzling, and a bed or group of them in bloom gives a magnificent effect to any lawn or garden.

True, Tall, Large-flowered Scarlet Gesneriana. Flowers of enormous size, on strong stems, the most durable of all Tulips, as it holds its beautiful color and keeps perfect several weeks. Color, rich crimson scarlet with glittering blue black centre. 3 for 10c., 35c. per doz.; or, buyer paying transit, $1.75 per 100.

Yellow Gesneriana. Similar in all respects to the Scarlet Gesneriana offered above, excepting in color, which is of a rich, bright effect ye yellow. 3 for 10c., 35c. per doz.; or, buyer paying transit, $1.75 per 100.

Rose Gesneriana. Blue based. 5c. each, 6 for 25c., 45c. per doz.

Rose Gesneriana. Blue and white based. 5c. each, 6 for 25c., 45c. per doz.

Rose Gesneriana. Pure white based. 5c. each, 6 for 25c., 45c. per doz.

Rose Flamed Gesneriana. (*See cut.*) 5c. each, 6 for 25c., 45c. per doz.

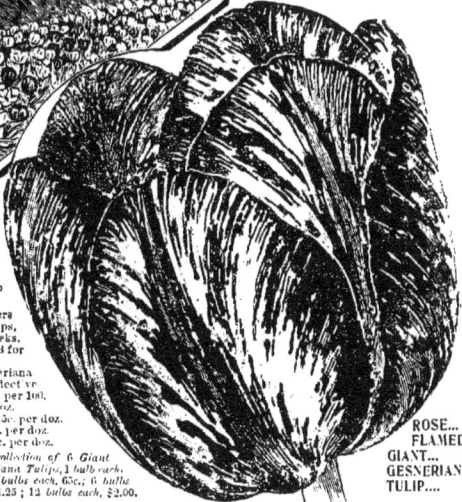

The collection of 6 Giant Gesneriana Tulips, 1 bulb each, 25c.; 3 bulbs each, 65c.; 6 bulbs each, $1.25; 12 bulbs each, $2.00.

ROSE... FLAMED GIANT... GESNERIANA TULIP....

NEW AND RARE NARCISSUS OR DAFFODILS...

The wonderful creations in new varieties have awakened an interest and enthusiasm among the lovers of flowers that has placed this, "**The Flower of the Poets,**" in the front rank of popularity, and they merit all the praise that can be bestowed upon them. Appearing, as they do, just after bleak winter, they turn our gardens into gorgeous masses of gold and silver, with a fragrance that is enchanting. They are equally valuable for growing in pots for winter flowering.

New Pure Yellow Phœnix Double Narcissus. The varieties "Silver" and "Orange Phœnix" are old favorite two-colored varieties. This new sort is of pure rich yellow. 20c. each, $2.00 per doz.

Daffodils with Large Long Trumpets.

Victoria. Immense bold erect flowers, with petals of remarkable breadth; color, creamy white with large broad frilled trumpet of rich yellow; strong grower and free bloomer. $2.00 each.

Hume's Giant Yellow. A very large and distinct Daffodil all yellow, the large trumpet being a shade darker. 35c. each, $3.50 per doz.

Daffodils with Cup-shaped Trumpets.

Maurice Vilmorin. Very large flower, with broad white petals; cup lemon, heavily stained orange-scarlet; dwarf. 60c. each, $5.00 per doz.

Beauty. A grand large flower, petals light yellow, with yellow bar; cup large, yellow, margined orange-scarlet; tall grower. 60c. each, $6.00 per doz.

Giant Poeticus. Nearly twice as large as the old favorite Poeticus; petals pure white, cup suffused orange; tall and handsome. $1.00 each, $10.00 per doz.

Cyclamen Flowered Daffodil.

Triandrus Albus, or Angel's Tears. A small but profuse-flowering sort, bearing several white Cyclamen-like flowers on a stem. A charming variety for pot culture. (*As the bulbs are small several should be put in a pot.*) 10c. each, 3 for 25c., 85c. per doz.

The Novelty Collection of Narcissus.

1 bulb each of the above for $4.00, delivered free in the United States.

Henderson's Bulb Culture.
Price, postpaid, 25c.; or given free as a Premium. See second page of cover.

Bulbs are delivered free in the United States. Avail yourself of our Premium offer on second page of cover.

... Double Amaryllis ...

(AMARYLLIS EQUESTRIS FL. PL.)

One of the most magnificent and gorgeous bulbous plants grown. The immense double flowers, rich light scarlet color with white shadings, and regal habit are simply incomparable. They throw up spikes from 18 inches to 2 feet high, bearing large double flowers of great substance, averaging 6 to 10 inches across. For pot culture in the window, conservatory or greenhouse they are well adapted, and when in bloom in the winter and spring months no flower can approach their beauty. (*See cut.*) (*Ready in November.*) **Price, $2.00 each.**

... Large Yellow Freesia ...

(F. LEICHTLINII MAJOR.)

Freesias are among our most popular bulbs for pot culture, flowering in the winter and spring in the conservatory or window garden. This large yellow-flowered variety is charming, with large primrose yellow flowers marked with orange blotches, and very fragrant. Six or eight bulbs should be planted in a 4-inch pot. The flowers are produced 6 to 8 on stems about 9 inches high, and are particularly useful for cutting, remaining in good condition for two weeks when kept in water.

Price, 5c. each, 6 for 25c., 45c. per doz., $3.00 per 100.

Red Calla ... (ARUM CORNUTUM.)

A very handsome Arum with red flowers spotted with black; stems curiously mottled green and white; foliage palm-like and very handsome; a showy pot plant for winter decoration.

Price, 25c. each, $2.50 per doz.

DOUBLE...
AMARYLLIS.

King of Snowdrops.

(GALANTHUS CASSABA-ROBUSTA.)

A new introduction and one of the best in the genus. The foliage is large, the growth robust and taller than others. The flowers are immense for a Snowdrop, having sepals over an inch long and rather broad, pure white, with intense dark green inner divisions; the flower stems are thick and upright and fine for cutting; the bulbs are large. In the early spring months there is nothing more beautiful than these snowy, graceful blossoms. Effects of surpassing beauty may be arranged with Snowdrops, mingling blue Scillas or Chionodoxas with them.

Price, 10c. each, 3 for 25c., 80c. per doz., $6.00 per 100.

Giant Spring Snowflake. (LEUCOJUM ... CARPATHICUM.)

This is similar to *Leucojum Vernum*, but much larger and decidedly more beautiful; the flowers, like monster snowdrops, often measure an inch across, are pure snow-white tipped with bright green. As the flowers open the yellow anthers are plainly shown, adding their charm to the exquisite flowers, which are borne in clusters on stout stalks about 9 inches high; the flowers are also deliciously fragrant. It is one of our earliest spring flowers, perfectly hardy, handsome in outline and prized for bouquets. When established, it produces enormous quantities of flowers. It can also be slowly forced in pots for winter bloom. (*See cut.*)   **Price, 15c. each, 3 for 40c., 6 for 75c., 12 for $1.25.**

Fortin's Giant...   Lily of the Valley.

This is the largest variety yet produced ; it is only adapted for open ground planting and may not show much superiority the first spring after planting, but by the second season, when it gets well established, it produces wonderfully luxuriant foliage and immense spikes crowded with purest white bells twice the size of any other sort. In every way it is a superior plant.

Price, large single crowns, 10c. each, 3 for 25c., 85c. per doz., $6.00 per 100.

Giant ....
Spring ...
Snowflake.

COPYRIGHT 1898 BY
PETER HENDERSON & CO

Japanese or "Pompadour Flowered" Double Herbaceous Pæonias.

These very novel and beautiful varieties produce double flowers with usually 3 sets of petals in tiers, one above the other, as shown in the illustration above.

"**Japanese Pride.**" (*Yamato-sangai.*) Lower tier of petals bright carmine rose; second tier of shorter ruffled petals are pure white surmounted by an upper tier of rose and blush.

"**Crimson Tower.**" (*Kamakurako.*) Very double crimson lake; the pompadour centre is composed of many wavy fluted petals.

"**Departing Sun.**" (*Benisangai.*) Lower tier of broad petals vermilion; second tier narrower white petals tipped rose; upper tier vermilion. (*See illustration on front cover.*)

"**Rare Brocade.**" (*Yayoura.*) Entire flower white shaded with cream at base of petals, outer edges of petals penciled with carmine.

Price for any of the above varieties, $1.50 each, or the set of 4 for $5.00, *delivered free in the United States.*

JAPANESE FILIFORM CENTRED HERBACEOUS PÆONIA.

COPYRIGHT 1898 BY PETER HENDERSON & CO

GIANT PINK "DOG'S-TOOTH VIOLET."

Giant Pink Dog's-Tooth Violet. (*Erythronium Johnsonii.*)

A new variety; one of the largest and handsomest of all Erythroniums; perfectly hardy. The foliage is charmingly variegated, and a mass of 15 or 20 plants is a pretty sight, even when not in flower, but when the graceful flowers are in bloom the effect is matchless. The plants luxuriate in rather moist, partially shady positions, and do very nicely when grown in pots in frames and brought into the conservatory or window garden for winter blooming. This variety bears from 6 to 12 rich rosy pink flowers, with orange centres on stems 12 to 18 inches high. (*See cut.*) 20c. each, $2.00 per doz.

Large Pink Wood Lily. (*Trillium Grandiflorum Roseum.*)

This is one of the largest and most beautiful of the American Wood Lilies, perfectly hardy, growing and flowering profusely in partially shady nooks about the lawn, under trees, etc. The flowers are large, often 5 inches across, of a bright rosy pink, changing as they age to deep wine red. If grown several in a pot it makes one of the best winter flowers. 20c. each, $2.00 per doz.

Sternbergia Macrantha.

A new and most charming addition to our autumnal flowering bulbs. The flowers, which are produced from September to November, are nearly 6 inches across when wide open under the influence of the bright sunlight, and of a bright golden yellow, slightly flaked with emerald, when they first open, much like a Crocus, but larger, and the petals more fleshy and of such firm texture that they withstand any amount of bad weather, brightening up our gardens long after other flowers are gone. They are not only very hardy but increase rapidly. 15c. each, $1.50 per doz.

Bulbs are delivered free in the United States, except where noted.
Note our Premium Offer on Second Page Cover.

SPIRÆA ASTILBOIDES FLORIBUNDA.

NEW AND RARE LILIES.

Ready in October.

Alexandræ—The White Auratum. This beautiful new variety produces pure white flowers, 5 to 6 inches long by 7 to 8 inches across; perfectly hardy, and of great merit. $10.00 per doz.

Auratum Platyphyllum. This is without a question one of the most wonderful Lilies in cultivation. The leaves are very long and broad, and the stems attain to a height varying from 7 to 10 feet. The flowers are similar in color to Auratum, heavily spotted, but are much larger, the petals more overlapping, and of greater substance. Immense bulbs. 60c. each, $6.00 per doz.

Alice Wilson. One of the most beautiful and distinct Lilies; the flowers, of a bright lemon yellow, are very large, borne erect and in clusters; very hardy, succeeds almost anywhere; height 1½ feet; very rare. $1.00 each, $10.00 per doz.

Nepalense. A new Lily from Nepaul. It is of great beauty, fragrant, large and attractive; flowers large, trumpet-shaped, of a bright sea-green, spotted vinous purple; very rare. $2.00 each.

Silver-edged St. Joseph's Lily (*L. longiflorum eximium foliis albo marginatis*). A Japanese variety of this well known hardy garden Lily, with healthy green foliage broadly striped with silvery white on the margins, a very unique and beautiful plant; the flowers are trumpet shaped, 6 to 8 inches long, fragrant, beautiful, snow-white. This is a splendid variety for forcing for winter flowers. Height, 2½ to 4 feet. In the open ground it blossoms in June and July. 50c. each, $5.00 per doz.

Wallichianum Superbum (or *Ochroleucum*). **The East India Yellow Lily.** A handsome new Lily introduced from India, attaining a height of from 4 to 8 feet. The flowers are funnel-shaped, slightly recurved and from 6 to 12 inches long. The color of the flower is most charming, the interior yellow, passing to white and often tinted pink; the exterior is yellow, flushed with purple in streaks. Sweetly perfumed. When planted in the open ground and well protected, it blooms in the later summer months; well adapted for pot culture. $2.00 each.

Maritimum. A very distinct and beautiful new species, growing 2 to 3 feet in height; the flowers are horizontal and bell-shaped, of a beautiful blood-crimson color, spotted maroon, on long wiry stalks. One of the most admired and certainly one of the most distinct of recent introductions. 50c. each, $5.00 per doz.

The Collection of New and Rare Lilies.

1 bulb each of the above-described 7 varieties, $6.50, delivered free in the United States.

SPIRÆA ASTILBOIDES FLORIBUNDA

This new variety is a great improvement over Spiræa Japonica. The flowers are borne in large, feathery panicles of purest white, and last a long time in bloom. It is dwarfer in habit, earlier and more profuse in bloom. This is undoubtedly one of the most beautiful and graceful subjects grown, its foliage and flowers combined making it one of the finest gems for winter and spring decoration in the house and conservatory, while for grace and elegance as a cut flower it is unequaled. Those who have tried it are enthusiastic in its praise. Fine clumps (ready in October), 30c. each, $3 00 per doz. (See cut.)

GIANT FLOWERING DAY LILY.

(Hemerocallis Aurantiaca Major.)

A new Hemerocallis from Japan; a very vigorous-growing plant with broad, long foliage, and immense broad-petaled trumpet lily-like flowers, 6 inches across, of a deep orange color, and very fragrant; it flowers freely during late summer and autumn. It is one of the best hardy perennials introduced in years, and has received highest awards from English horticultural societies. 60c. each, $6.00 per doz. (Ready in October.)

NEW DWARF CHINESE DAY LILY.

(Hemerocallis Middendorfii.)

A very distinct and rare variety which we are importing from China. The plant makes a dwarf, compact bushy growth only about 1 foot high, bearing large, lily-like flowers of dark orange yellow. 60c. each, $6.00 per doz. (Ready in October.)

NERINE FOTHERGILLI MAJOR.

This is one of the most beautiful bulbous plants for the conservatory or window garden. It is a vigorous grower, flowering with great certainty and producing clusters of large, wavy petaled, lily-like flowers of the most glittering vermilion scarlet. It blooms at various seasons. It requires potting but seldom, and should be left to grow, blossom and increase for several years, as well-established plants, when in flower, are simply magnificent. We know of no plant that will give more continued delight than this. 60c. each, $6.00 per doz.

THE "WHITE AURATUM."

LILIUM ALEXANDRÆ.

Bulbs delivered free in the United States, except where noted.   Avail yourself of our Premium offer on 2d page cover.

THE "WATER COLLECTION" OF HYACINTHS
FOR GLASSES

HYACINTHS grown in glasses are elegant and suitable ornaments for the parlor or sitting-room, and can be flowered in this way with very little trouble. The following assortment, which we have designated the "Water Collection," is made up of exceptionally beautiful varieties, such as we have found to succeed well when grown in glasses.

Directions for Growing Hyacinths in Glasses.

The bulbs should be placed in the glasses as early in the season as possible, keeping them in a cool, dark cellar or similar situation until the roots have nearly reached the bottom of the glass, after which the lightest and sunniest situation that can be had is the best; the water in the glasses should be changed two or three times a week, and in severe weather the bulbs must be removed from the window so as to be secure from frost. Fill the glass with water, so the base of the bulb will only touch the water.

SINGLES.

Bird of Paradise. Fine rich yellow, large compact spike.
General Pelissier. Rich bright crimson scarlet, splendid spike; very early and a great favorite for forcing, the best early dark red.
King of the Blues. Rich dark blue, the finest of the deep blues, extra large, well furnished, broad spike with large bells, erect and extra fine.
Lord Derby. Pale porcelain blue, grand large broad spike, splendid.
Moreno. Beautiful bright pink, large waxy bells, extra good truss, early.
Mr. Plimsoll. Ivory white tinted with rosy blush, grand spike of large bells.

DOUBLES.

Florence Nightingale. Pure glistening white, large double bells and splendid truss.
Frederick the Great. Bright rosy red, very large spike, semi-double.
Goethe. Rosy salmon with pinkish feather, fine bells and large truss, extra fine, the best of this color and very early.
Lord Wellington. One of the finest double blush pinks, large spike and bells, very early, extra
Murillo. Beautiful color, immense bells, azure blue, shaded darker, large strong spike, very early.
Othello. Deep black purple, large bells, very early.

COPYRIGHTED 1889 BY
PETER HENDERSON & CO
NEW YORK

Prices {delivered free in the United States}: The Water Collection of Hyacinths.
1 bulb each of the above 12 sorts—all separately named—$1.25, or we will supply any variety separately at 15c. each, $1.50 per doz.

THE "GEM" COLLECTION OF HYACINTHS.
FOR POT CULTURE

AN assortment of magnificent, superb and distinct varieties, of distinct shades of color and strong-growing, producing large and full spikes of bloom, and sure to give the greatest satisfaction for winter flowering in the conservatory or window garden, having been carefully selected from a great number of kinds.

Directions for Growing Hyacinths in Pots.

Fill a 5 or a 6 inch pot rather loosely to the brim with soil composed of loam, leaf mold and well-rotted cow manure and sand, in equal proportions; press the bulb down so that only one-fourth of it appears above the soil. Then water freely to settle the soil. The pots should then be plunged in a cold frame or in the garden, the top of the pots fully 2 inches below the surface of the ground, then cover with soil and leaves or litter to prevent freezing, this will encourage a strong growth of roots before the top starts. It must not be forgotten that a strong development of root can only be had at a low temperature, say from 40 to 50 degrees, and any attempt to force them to make roots quicker by placing them in a high temperature will certainly enfeeble the flower. There is no need to water, except at the time of potting, provided the pots have been covered up as directed. They may be removed in succession as required, after six to eight weeks' time, to a sunny window in a cool room.

SINGLES.

Cavaignac. Rosy carmine, shaded salmon, beautiful large broad bells, large compact spike.
Pink Charles Dickens. Waxy blush, suffused pink, tall spike, well filled, beautiful.
Grand Maitre. Grand large spike, pure ultramarine blue, shaded porcelain, strong grower.
King of the Blacks. Magnificent purplish black, compact truss.
La Grandesse. This is one of the finest pure white; grand compact spike, well furnished with very large bells.
Sonora. Orange tinged salmon, fine spike.

DOUBLES.

Double Charles Dickens. Fine dark blue, shaded lilac, large double bells and extra-large spike: the finest of the double dark blues.
La Grandeur. Bright golden yellow, with citron tinted centre, extra-large and double bells, straight strong truss.
Mad. Marmont. Unique color, white shaded azure, compact spike, extra fine bells.
Princess Louise. Superb deep red, large fine double bells and extra good spike.
The First. Bright rose pink, large bells and well furnished spike, extra fine, early.
Triumph Blandina. Peach-blossom pink, with faint stripe of carmine, tall, well-rounded spike; a lovely variety.

Prices {delivered free in the United States}: The Gem Collection of Hyacinths for Pots.
1 bulb each of the above 12 sorts—all separately named—$1.25, or we will supply any variety separately for 15c. each, $1.50 per doz.

CHOICE NAMED HYACINTHS

Single Hyacinth.

THESE grand Hyacinths hardly need special praise, as every one knows them to be the most useful and popular of hardly bulbs, and universal favorites in the most extended application of the word. They are not only largely grown for forcing into flower during the dull cheerless months of winter and early spring, but they are equally desirable for planting in beds or in the garden border. Hyacinths grown in glasses are elegant and suitable ornaments for the parlor or sitting-room, and can be flowered in this way with very little trouble. The single varieties are more generally used for this purpose.

The varieties prefixed * are the earliest flowering.

Hyacinth Bulbs at the single and dozen price are delivered free in the United States.

Dark Red.

SINGLES.

Amy, deep glossy carmine red, tall, well-filled spike. 10c. each, $1.00 doz.

Bazine. *See Novelties, page 5.*

***General Pelissier.** *See page 11.*

Robert Steiger, fine deep crimson, large truss. 10c. each, $1.00 doz.

***Veronica,** bright crimson, splendid color for bedding, and easily forced. 10c. each, $1.00 doz.

DOUBLES.

Baron von Pallandt, crimson, tipped green, fine bells and spike. 10c. each, $1.00 doz.

***Bouquet Tendre,** brilliant dark carmine red, good truss, early. 10c. each, $1.00 doz.

Lord Clarendon, dark red, large compact spike, fine bells. 12c. each, $1.25 doz.

***Louis Napoleon,** dark carmine red, rich and bright, fine spike, early. 12c. each, $1.25 doz.

Princess Louise. *See page 11.*

Rosy Red.

SINGLES.

Cavaignac. *See page 11.*

Fiancee Royal, very dark rose, fine color, large truss. 10c. each, $1 00 doz.

***Lord Macauley,** rosy carmine, large bells and spike, fine for pots and forcing. 12c. each, $1.25 doz.

Solfaterre, brilliant red, with orange yellow centre, fine spike. 10c. each, $1.00 doz.

***Von Schiller,** deep salmon pink, striped with crimson, large broad spike, forces easily. 12c. each, $1.25 doz.

DOUBLES.

***Alida Catherina,** very fine dark rose, good truss and very early. 10c. each, $1.00 doz.

Panorama, bright carmine rose, large bells, fine spike. 12c. each, $1.25 doz.

***Regina Victoria,** fine salmon rose, good bells and truss, early. 12c. each, $1.25 doz.

Frederick the Great. *See page 11.*

***Sir Joseph Paxton.** *See Novelties, page 5.*

Pink.

SINGLES.

***Gertrude,** rosy pink, large erect spike of exceptional value for pots or bedding, a splendid forcer. 10c. each, $1.00 doz.

Gigantea, rose and blush, large bells, extra large close truss, one of the best pinks for either forcing or bedding. 10c. each, $1.00 doz.

Maria Theresa, large, broad, waxy bells, blush pink, striped carmine. 10c. each, $1.00 doz.

***Moreno.** *See page 11.*

***Sultan's Favorite,** blush pink, widely striped carmine, large spike, early. 10c. each, $1.00 doz.

DOUBLES.

La Grand Concurrent, beautiful light rose, shaded with pink, grand spike and very large bells. 12c. each, $1.25 doz.

Koh-i-noor. *See Novelties, page 5.*

***Noble par Merite,** fine deep rose, large bells and extra good spike, one of the finest rose pinks, very early. 10c. each, $1.00 doz.

***Susanna Maria,** deep rose, large close spike, very early. 12c. each, $1.25 doz.

***The First.** *See page 11.*

Blush.

SINGLES.

Anna Paulowna, pretty rosy white, striped salmon, tall, fine spikes. 12c. each, $1.25 doz.

Cardinal Wiseman. *See Novelties, page 5.*

Pink Charles Dickens. *See page 11.*

***Norma,** delicate waxy blush pink with darker stripe, large bells, fine spike, one of the best pinks for forcing, very early and extra fine. 10c. each, $1.00 doz.

***Tubiflora,** splendid, immense bells, broad waxy petals, blush pink, shaded darker. 10c. each, $1.00 doz.

DOUBLES.

***Czar Nicholas,** delicate light pink, early. 10c. each, $1.00 doz.

***Groot Voorst,** fine rosy blush, large bells, fine truss. 10c. each, $1.00 doz.

***Pink Lord Wellington.** *See page 11.*

Prince of Orange, large, semi double light pink, fine spike. 12c. each, $1.25 doz.

Sir Walter Scott, delicate blush, shaded rosy pink, large double bells, first-class. 12c. each, $1.25 doz.

Pure White.

SINGLES.

Alba Superbissima, pure white, large, handsome spike. 10c. each, $1.00 doz.

***Baroness von Thuyll,** snow-white large compact truss of many bells, splendid for bedding and pot culture, one of the best for early forcing. 10c. each, $1.00 doz.

DOUBLES.

Couronne Blanche, pure white, fine bells and truss. 10c. each, $1.00 doz.

Duchesse de Bedford, pure white, large bells, extra. 10c. each, $1.00 doz.

Grand Vainqueur, pure waxy white, fine truss. 12c. each, $1.25 doz.

Pure White—Continued.

SINGLES.

*Grand Vedette, pure white, large bells, extra large spike, one of the best forcing varieties. 12c. each, $1.25 doz.
*La Grandesse. See page 11
*Reine de Holland, pure white, full spike, a favorite extra early variety. 10c. each, $1.00 doz.
Snowball. See Novelties, page 5.

DOUBLES.

Florence Nightingale. See page 11.
*La Tour d'Auvergne, the earliest pure white, extra fine, large bells and grand spike. 12c. each, $1.25 doz.
*Prince of Waterloo, fine, pure waxy white, fine truss of large bells, early. 12c. each, $1.25 doz.

Tinted White.

SINGLES.

Grandeur a Merveille, finest blush white, extra large truss; it is the best of this color to grow in quantity. 10c. each, $1.00 doz.
La Franchise, creamy white, waxy, large bells, good spike. 10c. each, $1.00 doz.
*Lord Grey, rosy white, very early, free-blooming. 12c. each, $1.25 doz.
Mr. Plimsoll. See page 11.
Semiramis, cream white, large waxy bells, fine spike. 12c. each, $1.25 doz.
Voltaire, pale blush white, fine spike. 10c. each, $1.00 doz.

DOUBLES.

Bouquet Royal, pure white, yellow centre, very double, fine spike, extra. 10c. doz.
*La Virginite, pale blush, rose centre, good bells, fine broad spike, very early. 10c. each, $1.00 doz.
*Mad. de Stael, blush white, rose centre, fine spike, very early. 12c. each, $1.25 doz.
Triumph Blandina. See page 11.
Venus. See Novelties, page 5.

Porcelain and Lavender.

SINGLES.

Captain Boynton. See Novelties, page 5.
Czar Peter, light porcelain, shaded lavender, large bells, grand spike, one of the finest of this color. 15c. each, $1.50 doz.
Lord Derby. See page 11.
*La Peyrouse, fine light blue, large bells, fine large spike; a splendid variety for bedding out, and a great favorite for forcing, very early. 10c. each, $1.00 doz.

DOUBLES.

*Blocksburg, light blue, marbled, extra large bells, magnificent large compact spike; the finest double light blue, early. 10c. each, $1.00 doz.
Mad. Marmont. See page 11.
*Pasquin, light lilac blue with dark centre, fine spike, early. 12c. each, $1.25 doz.
*Von Sieboldt, light porcelain, marked deep lavender, fine double bells, tall spike. 15c. each, $1.50 doz.

Bright Blue.

SINGLES.

Charles Dickens, bright blue, shading porcelain, excellent, large and compact spike; unexcelled for bedding or pot culture, and good for forcing. 10c. each, $1.00 doz.
Grand Maitre. See page 11.
*Leonidas, clear bright blue, handsome spike, early. 12c. each, $1.25 doz.
Queen of the Blues, the finest light blue, very large bells and tall fine spike. 15c. each, $1.50 doz.

DOUBLES.

Garrick, splendid bright blue, with dark centre, extra large truss, early. 12c. each, $1.25 doz.
Blue Lord Wellington, rich blue, striped with lilac, dark centre. 12c. each, $1.25 doz.
*Murillo. See page 11.
*Von Speyck. See Novelties, page 5.

Indigo and Purple.

SINGLES.

*Baron von Thuyll, deep violet blue, extra large compact spike, early and fine for forcing. 10c. each, $1.00 doz.
King of the Blues. See page 11.
Masterpiece. See Novelties, page 5.
Marie, deep purple blue, with light blue stripe, extra good compact and large truss; a splendid variety for pots. 10c. each, $1.00 doz.

DOUBLES.

Albion, dark purplish blue, large truss, late. 12c. each, $1.25 doz.
Bride of Lammermoor, dark purple, variegated centre, large bells, late. 12c. each, $1.25 doz.
Double Charles Dickens. See page 11.
*Prince of Saxe Weimar, dark blue, shaded violet, large truss of semi-double bells, early. 10c. each, $1.00 doz.

Black.

SINGLES.

*Baron von Humboldt, rich black violet, long spike. 12c. each, $1.25 doz.
King of the Blacks. See page 11.
Sir Henry Barckley, large black blue, good truss. 12c. each, $1.25 doz.

DOUBLES.

*Othello. See page 11.
Prince Albert, splendid dark black blue. 12c. each, $1.25 doz.
Robert Burns. See Novelties, page 5.

Yellow.

SINGLES.

Bird of Paradise. See page 11.
King of Yellows. See Novelties, page 5.
*Ida, pure bright yellow, large fine spike, the best pure yellow for early. 15c. each, $1.50 doz.
La Pluie d'Or, pale yellow, good truss, late. 10c. each, $1.00 doz.

DOUBLES.

Heroine, pure golden yellow, very fine large bells. 12c. each, $1.25 doz.
*Jaune Supreme, splendid rich yellow, tall stiff spike, well furnished with large bells, one of the yellowest, early. 12c. each, $1.25 doz.
La Grandeur. See page 11.
L'Or Vegetal, fine canary yellow with purplish feather, tall spike. 12c. each, $1.25 doz.

Orange and Apricot.

SINGLES.

Herman, fine orange yellow, large bells. 12c. each, $1.25 doz.
*King of Holland, rich orange, fine large spike and bells. 12c. each, $1.25 doz.
Prince of Orange, dark orange yellow. 10c. each, $1.00 doz.
Rhinoceros, orange, good truss. 12c. each, $1.25 doz.
Sonora. See page 11.

DOUBLES.

*Bouquet d'Orange, rosy salmon, fine semi-double bells, early. 12c. each, $1.25 doz.
*Goethe. See page 11.
Minerva. See Novelties, page 5.
William III., apricot, with pink centre, compact spike, very fine. 12c. each, $1.25 doz.

Hyacinth Bulbs at the single and dozen price are delivered free in the United States.

DOUBLE HYACINTH.

O UR MIXED HYACINTHS are good sized bulbs. (Cheap "scrubs" we do not handle.) Our bulbs average 6 inches in circumference and over. They are well adapted for open ground planting, and as we furnish them in separate colors, it enables the purchaser to plant them out as taste may dictate. Mixed Hyacinths, on account of their low price, are also extensively forced for winter blooming in the greenhouse and window garden, and for cutting purposes, but we strongly advise that Named Sorts be used when specimens are desired for pot and water culture. (See cut.)

Mixed Hyacinth Bulbs at the dozen price are delivered free in the United States, but at the 100 price purchaser pays transit.

MIXED SINGLE.	Doz.	Per 100	MIXED DOUBLE.	Doz.	Per 100
Dark red................	$0.60	$3.75	Dark red....................	$0.60	$3.75
Rose and pink...............	.60	3.75	Rose and pink...............	.60	3.75
Red and rose shades.......	.60	3.75	Red and rose shades........	.60	3.75
Pure white.	.60	3.75	Pure white..............	.60	3.75
Blush white and tints.......	.60	3.75	Creamy white and tints......	.60	3.75
Dark blue and violet........	.60	3.75	Dark blue and violet........	.60	3.75
Light blue................	.60	3.75	Light blue...............	.60	3.75
Blue, light and dark	.60	3.75	Blue, light and dark	.60	3.75
Yellow, all shades	.60	3.75	Yellow, all shades...........	.60	3.75
All colors...........	.55	3.50	All colors..............	.55	3.50

3 sold at dozen rates ; 25 sold at 100 rates. 1,000 rates quoted on application.

Henderson's Special Bedding and Forcing Hyacinths

The bulbs under this heading are of a much larger size than those put in mixtures; in fact, most of these average 7 inches in circumference, or as large as those often sent out as first size by some dealers. They are of the same age as the first size Fancy picked bulbs offered by us on pages 5, 11, 12 and 13, from which they are the second selection. They will produce grand spikes of bloom, and the different sorts, as offered below, all being of one shade of color and all blooming at one time, render them of great value for forcing or bedding out, especially in designs, enabling one to secure the exact color effect desired, which is a great advantage over mixed reds, mixed blues, etc., at but a slight advance in price.

Henderson's Special Bedding and Forcing Hyacinth Bulbs at the dozen price are delivered free in the United States, but at the 100 price purchaser pays transit.

3 bulbs supplied at the dozen rate ; 25 at the 100 rate.

PRICES OF HENDERSON'S SPECIAL BEDDING AND FORCING HYACINTHS.

SINGLE VARIETIES.	Per doz.	Per 100	DOUBLE VARIETIES.	Per doz.	Per 100
Single Brilliant crimson..................................	$0.75	$5.00	Double Brilliant crimson........................	$0.75	$5.00
" Rosy red..	.75	5.00	" Rosy red..	.75	5.00
" Pink..	.75	5.00	" Pink..	.75	5.00
" Blush	.75	5.00	" Blush...	.75	5.00
" Snow white...	.75	5.00	" Snow white......................................	.75	5.00
" Ivory white...	.75	5.00	" Ivory white......................................	.75	5.00
" Lavender...	.75	5.00	" Lavender..	.75	5.00
" Bright blue...	.75	5.00	" Bright blue......................................	.75	5.00
" Purple...	.75	5.00	" Purple..	.75	5.00
" Yellow	.75	5.00	" Yellow..	.75	5.00
" Orange	.75	5.00	" Orange..	.75	5.00
" Special Mixed.....................	.65	4.50	" Special Mixed.................	.65	4.50

COLLECTIONS OF
HYACINTHS FOR BEDS.

HYACINTH BED "A."

HYACINTH BED "B."

FOR convenience of those who do not quite understand arranging beds, we have made a selection of sets at prices much below the regular rates, to encourage more extensive planting this autumn. The bulbs we supply for these beds are selected from Henderson's Special Bedding and Forcing Hyacinths offered on page 14. Should you prefer other combinations of color than those we have arranged below, then make your selection from colors offered on lower half of page 14, and we will furnish the beds at the same prices as offered below. Plant the bulbs 6 inches apart.

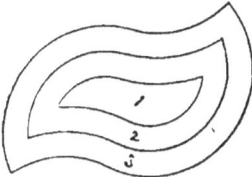

HYACINTH BED "A."

6 feet long by 3 feet across; requires a total of 75 Hyacinth bulbs planted 6 inches apart. **Price** for either of below combinations of colors, $3.50, purchaser paying transit.

Crimson, White and Blue Combination for Bed "A."
No. 1 requires 20 crimson Hyacinths; No. 2, 25 white; No. 3, 30 blue.

Lavender, White and Pink Combination for Bed "A."
No. 1 requires 20 lavender Hyacinths; No. 2, 25 white; No. 3, 30 pink.

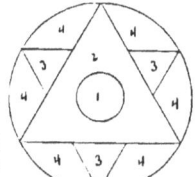

HYACINTH BED "B."

15 feet in circumference or 5 feet across; requires a total of 100 Hyacinth bulbs planted 6 inches apart. **Price** for either of below combinations of colors, $4.50, purchaser paying transit.

Crimson, White, Blue and Rose Combination for Bed "B."
No. 1 requires 9 crimson Hyacinths; No. 2, 30 white; No. 3, 21 blue; No. 4, 40 rose.

Purple, Pink, Lavender and White Combina'n for Bed "B."
No. 1 requires 9 purple Hyacinths; No. 2, 30 pink; No. 3, 21 lavender, No. 4, 40 white.

HYACINTH BED "C."   HYACINTH BED "C."

HYACINTH BED "C."

15 feet in circumference or 5 feet across, requiring a total of 100 bulbs planted 6 inches apart. **Price** for either of below combinations of colors $4.50, purchaser paying transit.

Rose, Yellow, Blue, White and Crimson Combination for Bed "C."
1st or centre row requires 9 rose Hyacinths; 2d row, 14 yellow; 3d row, 20 blue; 4th row, 25 white; 5th or outside row, 32 crimson.

Lavender, White, Crimson, Pink and Purple Combination for Bed "C."
1st or centre row requires 9 lavender Hyacinths; 2d row, 14 white; 3d row, 20 crimson; 4th row, 25 pink; 5th or outside row, 32 purple.

HYACINTH BED "D."

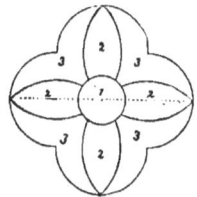

15 feet in circumference or 5 feet across; requires a total of 100 bulbs planted 6 inches apart. **Price** for either of below combinations of colors, $4.50, purchaser paying transit.

Blue, White and Crimson Combination for Bed "D."
No. 1 requires 32 purple Hyacinths; No. 2, 32 white; No. 3, 36 crimson.

Pink, Lavender and White Combination for Bed "D."
No. 1 requires 32 pink Hyacinths; No. 2, 32 lavender; No. 3, 36 white.

HYACINTH BED "E."

6 feet across; requires a total of 120 bulbs planted 6 inches apart. **Price** for either of the below combinations of colors, $5.25, purchaser paying transit.

Lavender, Pink and White Combination for Bed "E."
No. 1 requires 10 lavender Hyacinths; No. 2, 50 pink; No. 3, 60 white.

White, Crimson and Purple Combination for Bed "E."
No. 1 requires 10 white Hyacinths; No. 2, 50 crimson; No. 3, 60 purple.

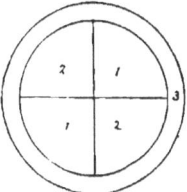

HYACINTH BED "D."   HYACINTH BED "E."

ROMAN HYACINTHS

THE flowers while smaller than those of the ordinary Dutch Hyacinth, yet are produced in much greater abundance, each bulb producing several graceful spikes of bloom; their delicious perfume, earliness and profusion of bloom have made Roman Hyacinths exceedingly popular. They are so easily grown and so early that if potted in August and September they can be brought into flower in November and December, and by a succession of plantings can be had in bloom throughout the winter. Roman Hyacinths are not considered hardy enough for open ground culture north of Washington, though in our grounds near New York they have proved perfectly hardy, slightly protected. For cutting purposes the Roman Hyacinth is the finest bulb grown. (See cut.)

WHITE ROMAN HYACINTH.

PRICES OF ROMAN HYACINTHS.

Delivered free in the U. S. at the single and dozen price, but at the 100 rate purchaser pays transit.

White Romans. Select Bulbs	5c. each,	50c. per doz.,	$3.50 per 100.
Double White Romans.	10c. each,	85c. per doz.,	$5.00 per 100.
Canary Yellow Romans.	10c. each,	88c. per doz.,	$5.00 per 100.
Blush Pink Romans.	5c. each,	40c. per doz.,	$2.50 per 100.
Dark Rose Romans.	5c. each,	40c. per doz.,	$2.50 per 100.
Light Blue Romans.	5c. each,	40c. per doz.,	$2.50 per 100.

WHITE ITALIAN OR PARISIAN HYACINTHS.

These are extensively grown by florists for the flower markets, as the profusion of bloom, fragrance and graceful spikes render them very popular for cutting purposes. They flower quickly after being potted and deserve to be more largely grown. The bulbs are red-skinned and bloom a little later than White Romans, and therefore are valuable for succession; valuable for outside planting for cutting purposes in the spring.

Price, 5c. each, 50c. per doz., $3.00 per 100.

Delivered free in the United States at the single and dozen price, but at the 100 rate purchaser pays transit.

PAN HYACINTHS.

The Hollanders practice this method of growing Hyacinths for winter flowering in the house with most satisfactory results: 10 to 12 bulbs are planted in earthen pans, 8 or 9 inches across. Several of these pans can be filled and put away to root in the usual manner, and by bringing them in at intervals a beautiful display can be enjoyed for weeks. Bulbs of one variety are usually preferred because they will all flower at the same time, though, if a person prefers a variety of colors in one pan, selections can be made from the undermentioned colors which all bloom at about the same time, under the same growing conditions.

Pure White, Tinted White, Pink, Crimson, Light Blue, Dark Blue or Mixed Colors.

Price, 5c. each, 50c. per doz., delivered free in the United States $2.75 per 100, purchaser paying transit.

A GROUP OF GRAPE HYACINTHS.

"PAN" HYACINTHS.

GRAPE, MUSK AND FEATHERED HYACINTHS.

Are very pretty for permanent beds and edgings and partially shaded situations; they should be planted in groups of one dozen or more, where they will soon spread, and being perfectly hardy will take care of themselves. They are very pretty when grown six in a pot for winter flowering in the house.

Grape Hyacinths. Produce flowering spikes about 6 inches high, with little round bells so arranged as to resemble a bunch of grapes.
" " **Blue.** 2 bulbs for 5c., 15c. per doz., $1.00 per 100.
" " **White.** 3 bulbs for 10c., 25c. per doz., $1.80 per 100.
Musk Hyacinths. A small growing Hyacinth of a purplish color, emitting, when in flower, a strong and very agreeable musky odor. 15c. each, $1.50 per doz.
Feathered Hyacinths. *Feathery, plume-like* spikes, 9 to 12 inches high. Striking in the garden and useful for cutting. 3 bulbs for 10c., 25c. per doz., $1.50 per 100.

6 sold at dozen rates, 25 at 100 rates, 250 at 1,000 rates. Delivered free in the United States, except where noted.

Collections of Single Early Tulips

... FOR BEDS. ...

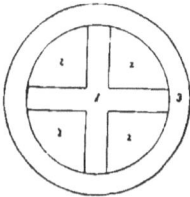

Bed "G."

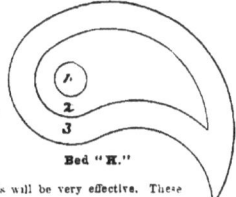

Bed "H."

We offer these sets at prices much below the regular rates to encourage a large number of our customers to put out beds of Tulips this autumn, knowing they will give unbounded satisfaction.

For beauty of form and brilliancy of color nothing can equal the gorgeous effects of Tulips. They are perfectly hardy and of the easiest culture. The following "**Collections of Tulips for Beds**" we have made up for the special convenience of those who do not understand the selections of sorts that bloom simultaneously and are of equal heights. We have selected bright contrasting colors, and the beds will be very effective. These Collections can only be sent by express or freight, buyer paying transit.

BED OF SINGLE EARLY TULIPS No. "G."

This bed is 18 feet in circumference, or 6 feet across, and requires a total of 200 bulbs planted 5 inches apart. **Price**, for either of below color combinations, $3.00, *buyer paying transit.*

White, Scarlet and Yellow Combination for Bed "G."— No. 1 requires 40 white Tulips; No. 2, 80 scarlet; No. 3, 80 yellow.

Crimson, White and Pink Combination for Bed "G."—No. 1 requires 40 crimson Tulips; No. 2, 80 white; No. 3, 80 pink.

BED OF SINGLE EARLY TULIPS No. "H."

This bed is 10 feet long by 6 feet across at the widest part, and requires a total of 275 bulbs planted 5 inches apart. **Price**, for either of below color combinations, $3.75, *buyer paying transit.*

Scarlet, Yellow and Pink Combination for Bed "H."—No. 1 requires 25 scarlet Tulips; No. 2, 100 yellow; No. 3, 125 pink.

Yellow, Crimson and White Combination for Bed "H."—No. 1 requires 25 yellow Tulips; No. 2, 100 crimson; No. 3, 150 white.

Bed "J."

BED OF SINGLE EARLY TULIPS No. "J."

This bed is 18 feet in circumference, or 6 feet across, and requires a total of 200 bulbs planted 5 inches apart. **Price**, for either of below color combinations, $3.00, *buyer paying transit.*

Yellow and Scarlet Combination for Bed "J."—Nos. 1 and 3 require 50 yellow Tulips each; Nos. 2 and 4, 50 scarlet each.

White, Crimson, Yellow and Pink Combination for Bed "J."—No. 1 requires 50 white Tulips; No. 2, 50 crimson; No. 3, 50 yellow; No. 4, 50 pink.

BED OF SINGLE EARLY TULIPS No. "K."

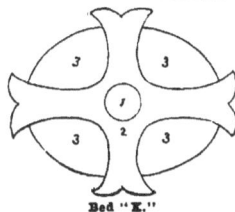

Bed "K."

This bed is 6½ feet long by 5 feet across, and requires a total of 215 bulbs planted 5 inches apart. **Price**, for either of the below color combinations, $3.50, *buyer paying transit.*

White, Scarlet and Yellow Combination for Bed "K."—No. 1 requires 15 white Tulips; No. 2, 120 scarlet; No. 3, 80 yellow.

Crimson, White and Pink Combination for Bed "K."—No. 1 requires 15 crimson Tulips; No. 2, 120 white; No. 3, 80 pink.

BED OF SINGLE EARLY TULIPS No. "L."

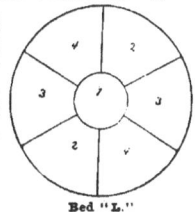

Bed "L."

This bed is 18 feet in circumference, or 6 feet across, and requires a total of 200 bulbs planted 5 inches apart. **Price**, for either of the color combinations given below, $3.00, *buyer paying transit.*

Scarlet, Yellow, White and Rose Combination for Bed "L."—No. 1 requires 20 scarlet Tulips; the two No. 2s, 60 yellow; the two No. 3s, 60 white; the two No. 4s, 60 rose.

Yellow, Crimson and White Combination for Bed "L."—No. 1 requires 20 yellow Tulips; and the 6 sections 30 Tulips each, alternately crimson, then white.

DUC VAN THOL TULIPS.

TULIPS are such universal favorites that it is scarcely necessary to expatiate upon their merits here. Their ease of culture, combined with beauty of form and gorgeous coloring, renders them the most popular bulbs grown for spring bedding, and for winter flowering in the greenhouse and window garden they are incomparable. The Tulip is extremely hardy and of easy culture, flowering as freely in the shade as in the sunshine, and producing as fine flowers in a confined town garden as in a more favored place. Double and Single Tulips, when associated together and planted in front of shrubs, maintain a longer display than if either are separately planted. In gardens where the flower beds are to be kept gay from the earliest day of spring, plant between the lines of Tulips, Scilla Sibirica, Chionodoxas, Snowdrops or Crocus, as these flower first and are through when the Tulips come into bloom.

The letters A, B, C, following the varieties, indicate their earliness of bloom; the A's flower together and are the earliest; B next, etc. The average height in inches is also given to aid in arranging flower beds.

Tulip Bulbs are delivered in the United States FREE at the single and dozen price, but at the 100 and 1,000 price purchaser pays transit.

DARK BLOOD RED.

Bacchus. *A* 10. Dark, blackish scarlet, large, well-formed flower, large yellow base; early, good forcer. 3 for 10c., 30c. per doz., $1.50 per 100.
De Keyser. *B* 9. A large, magnificent dark scarlet, with golden base. 3 for 10c., 35c. per doz., $2.00 per 100.
Koh-i-noor. *B* 7. Very rich, dark blood-red, grand. 5c. each, 50c. per doz., $2.75 per 100.
Purple Crown (d'Admiration). *A* 8. Rich blackish crimson, yellow base. 2 for 5c., 25c. per doz., $1.25 per 100.

BRIGHT CRIMSON SCARLETS.

Artus (Garibaldi). *B* 7. Dwarf, bright red, fine bold flowers. 2 for 5c., 25c. per doz., $1.25 per 100.
Belle Alliance (Waterloo). *B* 9. Brilliant scarlet, yellow base, large flower, fine for bedding. 3 for 10c., 70c. per doz., $1.75 per 100.
Couleur Cardinal. *C* 10. Distinct and fine, large flowers, orange scarlet. 3 for 10c., 35c. per doz., $2.00 per 100.
Crimson King (Roi Cramoisi). *B* 9. Fine showy bedder, large flowers, bright crimson, yellow base. 2 for 5c., 25c. per doz., $1.25 per 100.
Pottebakker, scarlet. *A* 9. Bright scarlet, very large and fine flower. 3 for 12c., 40c. per doz., $2.25 per 100.
Rembrandt. *A* 9. Superb, rich glowing crimson, large flower, early and fine forcer. 3 for 10c., 35c. per doz., $2.00 per 100.
Verboom. *B* 9. Bright scarlet, large flower. 3 for 10c., 35c. per doz., $2.00 per 100.
Vermilion Brilliante. *B* 5. Deep, dazzling vermilion, magnificent, extra fine for pots and massing, very effective. 5c. each, 50c. per doz., $3.00 per 100.

ROSY REDS.

Adeline. *B* 8. A lovely and large Tulip of a splendid satiny cherry pink, with yellow base, dwarf and fine for bedding. 3 for 12c., 45c. per doz., $2.50 per 100.
Couleur Ponceau. *C* 10. Rosy crimson, large white base. 2 for 5c., 25c. per doz., $1.25 per 100.
Proserpine. *B* 12. The "Queen of Tulips," large and very effective, rich, silky, carmine rose of perfect form. 5c. each, 50c. per doz., $3.50 per 100.
Stanley. *B* 10. Beautiful rosy carmine, large flower. 3 for 12c., 45c. per doz., $2.50 per 100.
Vander Helst. *A* 10. Extra large dark rosy crimson, white base. 3 for 12c., 35c. per doz., $2.00 per 100.

ORANGE REDS.

Brutus. *B* 9. Bright, orange crimson, with small golden margin, very showy. 3 for 10c., 30c. per doz., $1.75 per 100.
Duchesse de Parma. *B* 9. Orange red, banded yellow, very large fine flower, early. 3 for 10c., 30c. per doz., $1.50 per 100.
Prince of Austria. *B* 9. Glittering copper red, large flower, fine forcer, sweet scented. 5c. each, 50c. per doz., $3.50 per 100.
Thomas Moore. *B* 10. Orange, effective large Tulip, sweet-scented. 3 for 12c., 45c. per doz., $2.50 per 100.

RED WITH YELLOW EDGE.

Duc de Berlin. *A* 8. Brilliant light scarlet, broadly edged yellow, and yellow base, early. 3 for 10c., 35c. per doz., $2.00 per 100.
Kaiser-Kroon. *B* 10. Deep crimson, with broad, orange-yellow margin, large and exceedingly effective. 3 for 10c., 35c. per doz., $2.25 per 100.

CARMINE ROSE.

Cottage Maid (La Precieuse). *B* 9. Carmine pink, centre of petals feathered white, base yellow, a charming variety. 5c. each, 50c. per doz., $4.00 per 100.
Rose Luisante. *C* 8. Dark rose, extra fine. 6c. each, 60c. per doz., $4.00 per 100.

BLUSH PINKS.

Princess Marianne. *B* 9. Peach blossom pink, suffused darker pink, yellow base, extra large flower. 3 for 12c., 45c. per doz., $2.50 per 100.
Rosa Mundi Huyckman. *B* 9. Carmine pink, feathered white through centre of petals, white base, large flower. 3 for 12c., 45c. per doz., $2.50 per 100.
Rose Grisdelin. *B* 6. Delicate rose, shaded white, fine for forcing, very beautiful. 5c. each, 50c. per doz., $2.75 per 100.

CLARET PURPLES.

Queen of the Violets (Pres. Lincoln). *B* 9. Rosy claret, large white base, large fine flower. 3 for 10c., 35c. per doz., $2.00 per 100.
Vander Neer. *C* 8. Rich claret purple, large fine flower, the finest of this class. 3 for 10c., 30c. per doz., $1.75 per 100.
Wouverman. *C* 8. Dark claret violet, very large flower, beautiful color. 3 for 10c., 30c. per doz., $1.50 per 100.

YELLOWS.

Canary Bird. *A* 8. Clear, rich golden yellow, excellent for either early forcing or garden. 3 for 12c., 40c. per doz., $2.25 per 100.
Chrysolora. *B* 9. Pure yellow, large and handsome flower. 3 for 12c., 40c. per doz., $2.25 per 100.
Mon Tresor. *B* 9. Grand, rich yellow, large fine flower, 4 inches across. 5c. each, 50c. per doz., $2.75 per 100.
Pottebakker, yellow. *A* 9. Large, fine bright yellow, feathered crimson. 5c. each, 50c. per doz., $2.75 per 100.
Yellow Prince (La Pluie d'Or). *B* 9. Bright yellow, large and showy, the finest low-priced yellow for bedding out. 3 for 12c., 40c. per doz., $2.25 per 100.

WHITE.

Joost Van Vondel, white. *B* 10. Silvery white, large and of fine form, the finest white Tulip. 15c. each, $1.50 per doz.
La Reine (Queen Victoria). *B* 8. White, very faintly tinted rose, good for both bedding and forcing. 2 for 5c., 25c. per doz., $1.25 per 100.
L'Immaculee. *B* 9. Pure white, yellow base, large broad petaled flower, very early and fine. 3 for 10c., 30c. per doz., $1.50 per 100.
Pax Alba. *B* 8. Large, pure white flowers, with yellow base. 3 for 10c., 30c. per doz., $1.75 per 100.
Pottebakker, white. *A* 9. Pure white, fine, large and showy flower, excellent forcer. 5c. each, 50c. per doz., $3.00 per 100.

6 bulbs of one variety sold at dozen rates, 25 at 100 rates. 1,000 rates on application.

SINGLE EARLY FLOWERING TULIPS—Continued.

VARIEGATED YELLOW AND RED.

Golden Bride of Haarlem. *B* 8. Golden yellow, feathered bright crimson, extra fine and beautiful. 3 for 12c., 45c. per doz., $2.50 per 100.
Golden Standard. *B* 8. Bright red with golden stripes, extra fine. 3 for 10c., 35c. per doz., $2.00 per 100.
Marquis de Westrade. *B* 9. Deep golden yellow, with a ¼ inch stripe of dark blood red from base to tip of each petal. 5c. each, 50c. per doz., $3.00 per 100.

VARIEGATED WHITE AND RED.

Comte de Vergennes. *C* 12. Large flower of rich crimson, flaked with pure white. 3 for 12c., 45c. per doz., $2.50 per 100.
Grand Duke de Russie. *B* 9. (Fabiola.) A fine, large and distinct Tulip; rosy purple, flaked and striped with carnation and white. 3 for 12c., 40c. per doz., $2.25 per 100.
Grand Master of Malta. *B* 9. Dark crimson, feathered white, magnificent. 3 for 12c., 40c. per doz., $2.25 per 100.
Globe de Rigaut. *B* 10. Large, broad flower, violet, base white, zoned indigo. 3 for 12c., 40c. per doz., $2.25 per 100.
Joost Van Vondel. *B* 10. Deep cherry red, with white feather through centre of petals, beautiful. 3 for 12c., 45c. per doz., $2.50 per 100.
La Precieuse Rectified. *B* 8. Very large, broad petaled flower, white feathered and picotee edged with light and dark pink, yellow base. 3 for 12c., 45c. per doz., $2.50 per 100.
Silver Standard (Royal Standard.) *B* 8. White, feathered with rosy crimson. 3 for 10c., 35c. per doz., $2.00 per 100.

DUC VAN THOL TULIPS.

Very early and especially valuable for forcing and pot culture. Height from 6 to 7 inches. (See cut.)

Red and Yellow. 2 for 5c., 25c. per doz., $1.25 per 100.
Crimson (carmine). 2 for 5c., 25c. per doz., $1.25 per 100.
White. 3 for 12c., 40c. per doz., $2.25 per 100.
Yellow. 5c. each, 50c. per doz., $3.00 per 100.
Rose. 3 for 12c., 40c. per doz., $2.25 per 100.
Scarlet. Rich dazzling scarlet, highly prized as an early forcing sort and for pot culture. 2 for 5c., 25c. per doz., $1.25 per 100.
Gold Laced. Gold and crimson striped. 3 for 10c., 30c. per doz., $1 50 per 100.
Claret, edged white. Fine large flower. 3 for 10c., 30c. per doz., $1.75 per 100.
Orange. Reddish orange. 3 for 10c., 30c. per doz., $1.75 per 100.

Collections of the 60 named Single Early Tulips.

Offered on pages 18 and 19 (exclusive of the "Odd and Novel" Tulips offered opposite).

60 bulbs, 1 bulb each of the 60 varieties, $2.25.
180 bulbs, 3 of the 60 varieties, $6.00.
360 bulbs, 6 of the 60 varieties, $10.00.
Delivered free in the U. S.

WHITE POTTEBAKKER.   COTTAGE MAID.

Single Early Flowering Mixed Tulips.

First quality, 20c. per doz., delivered in the U. S., or, buyer paying transit, for 75c. per 100, $6.00 per 1,000.

Henderson's Special Mixture. Made up from named sorts, proper proportions of bright colors, all blooming together and of uniform height. 25c. per doz., $1.00 per 100, $8.50 per 1,000.

Various Odd and Novel Early Flower= ing Tulips.

Clusiana Major or Lady Tulip. Delicately tinted white, flushed with red on the outside, with a conspicuous purple-black base. 10c. each, $1.00 per doz.
Greigii. The "Queen of Tulips." An exceedingly handsome and distinct species. Flowers very large, of brilliant orange-scarlet, with a yellow and black centre. The foliage is oddly spotted with dark maroon. 20c. each, $2.00 per doz.
Oculis Solis, or Sun's Eye Tulip. A very novel, beautiful and showy variety, dazzling, fiery red, with black eye. 5c. each, 50c. per doz.
Large-flowering Sweet-scented Florentine. A fine variety for either pots or garden. Color, yellow. The flowers emit the odor of violets. 5c. each, 50c. per doz.
Cornuta. (Chinese Horned Tulip.) Yellow, striped red, curiously twisted petals like spiral horns. 5c. each, 50c. per doz.
Retroflexa. Large clear yellow flower, petals long-pointed and gracefully recurved. 5c each, 50c. per doz.
Viridiflora Præcox. Novel and beautiful, immense flowers of green and white. 5c. each, 50c. per doz.

The Collection of "Various Odd & Novel" Tulips.

Delivered free in the U. S.

7 bulbs, 1 each of above 7 varieties, $0.50
21 bulbs, 3 each of above 7 varieties, 1.25
42 bulbs, 6 each of above 7 varieties, 2.25

A USEFUL PAMPHLET.

BULB CULTURE

BY PETER HENDERSON.

Bulbs, alphabetically arranged, with Special Cultural Instructions for each—Designs for Beds of Tulips and Hyacinths—Preparation and Outside Planting—Spring Flowering Bulbs—Winter Flowering Bulbs—Summer Flowering Bulbs—Forcing Bulbs—Naturalizing Bulbs in Lawns, etc.—The Window Garden of Bulbs—Bulbs in Cold Frames and Pits—When Bulbs should be Taken Up, etc., etc.

24 pages. Price, postpaid, 25 cents, or allowed as premium on an order. See 3d page of cover.

6 bulbs of one variety sold at dozen rates, 25 at 100 rates. Delivered free in U. S. except where noted.

LATE MAY FLOWERING GARDEN TULIPS.

BIZARRE TULIP.

THIS group of Late or May-flowering Garden Tulips generally are in full flush of bloom about "Decoration Day." They differ from the Early Single Tulips by their taller growth and later bloom; planted in conjunction with the early Tulips—on pages 18 and 19—a gorgeous display of bloom can be enjoyed until late in May. These late Tulips are brilliant in the extreme for bedding purposes. Their flowers are very large, symmetrically formed, and their magnificent colors, with interesting and delicate featherings and markings, make a sight not soon to be forgotten, particularly when bedded out in quantity. They are very hardy and do well through all sorts of weather.

❧❧ BIZARRE TULIPS. ❧❧

Grand rich flowers of perfect shape, having yellow ground color, feathered or striped with crimson, purple or white.

Choice Mixed Bizarres. 3 for 10c.; 30c. per doz., *delivered in the United States;* or $1.50 per 100, *buyer paying transit.*

Everet Krosschell. A magnificent and distinct Tulip; the leaves are 5 inches wide, flower immense, individual petals are three inches across, the coloring is beautiful, and intricately veined and blotched combination of gold, canary yellow, orange red and maroon, satiny and rich.

Gouden Munt. Dark glossy red, broadly feathered-edged, with rich yellow base, markings sharp and distinct, very showy.

La Citadel. Light yellow, picotee edged, and flamed throughout with chestnut and violet.

Fenelon. Golden brown, feathered violet, mahogany and yellow, large yellow base.

Cortez. Deep yellow, overlaid with featherings of chestnut red and violet.

Adeline Patti. Rich deep yellow, picotee edged, and flaked mahogany red.

Price of any of the above-named Bizarres, 5c. each, 50c. per doz., *delivered in the United States;* or $3.00 per 100, *buyer paying transit.*

The collection of six, 1 bulb each, 25c.; 3 bulbs each, 70c.; 6 bulbs each, $1.25; 12 bulbs each, $2.00, *delivered in the United States.*

VIOLET BYBLŒMEN TULIPS.

Blotched, striped or feathered with blue, lilac, violet, purple or black on white ground.

Choice Mixed Violet Byblœmens. 3 for 10c., 30c. per doz., *delivered in United States;* or $1.75 per 100, *buyer paying transit.*

La Grande Duchesse. Purple, maroon, claret and heliotrope, white feathering.

Brunhilde. White ground, lightly splashed purple, picotee, edged with violet.

Graf von Buren. Light violet ground, feathered with heliotrope and bright red, and splashed white.

Paul Kruger. Light wine red, suffused carmine, white blotches, at intervals.

Potgieter. White, veiled lavender, with splashes of violet, edges feathered violet.

Van 1st. Black purple, suffused violet, feathered white.

Price of any of the above-named Violet Byblœmens, 5c. each, 50c. per doz., *delivered in the United States;* or $3.00 per 100, *buyer paying transit.*

The collection of six named sorts, 1 bulb each, 25c.; 3 bulbs each, 70c.; 6 bulbs each, $1.25; 12 bulbs each, $2.00, *delivered in United States.*

ROSE BYBLŒMEN TULIPS.

Magnificent flowers, perfection in form, having a white ground color with beautiful stripes and markings of crimson, pink and scarlet and rose.

Choice Mixed Rose Byblœmens. 3 for 10c., 30c. per doz., *delivered in United States;* or $1.75 per 100, *buyer paying transit.*

L'Estemei. White ground heavily feathered, bright red, and blotched lightly with claret and red.

Rembrandt. White pink, ground heavily suffused with bright red.

Gen'l Gurko. A claret and bright carmine stripe throughout each petal, feathers out over a white ground, margin white, base of flower indigo.

Proteus. White ground, flaked rosy carmine.

Vondel. Rosy ground, striped dark red, irregular white blotches.

Phœnix. Bright fiery crimson scarlet, feathered white and pink in centre of petals.

Prices. Any of the above-named Rose Byblœmens, 5c. each, 50c. per doz., *delivered in United States;* or $3.00 per 100, *buyer paying transit.*

The collection of six named sorts, 1 bulb each, 25c.; 3 bulbs each, 70c.; 6 bulbs each, $1.25; 12 bulbs each, $2.00, *delivered in United States.*

PICOTEE EDGED TULIPS.

Maiden's Blush. An elegant, long-shaped, clear white flower; the petals, which are pointed and elegantly reflexed, are beautifully margined and penciled on the edges with bright pink. 5c. each, 50c. per doz., *delivered in United States;* or $3.00 per 100, *buyer paying transit.*

Golden Eagle. (Golden Crown.) Large, showy, butter-yellow flowers, with pointed petals, each petal edged crimson. 3 for 10c., 30c. per doz., *delivered in United States;* or $2.00 per 100, *buyer paying transit.*

PICOTEE EDGED TULIP.

All Bulbs delivered free in the United States, except where noted.

PARROT TULIPS.

PARROT or ... DRAGON TULIPS.

These belong to the late or May-flowering Tulips, and have immense attractive flowers of singular and picturesque forms and brilliant and varied colors. The petals are curiously fringed or cut, and the form of the flower, especially before it opens, resembles the neck of a Parrot. They form extravagantly showy flower beds, an endless variety of form and color, and should be grown in every flower-garden in quantities.

Parrot Tulips, Mixed Colors. 2 for 5c., 25c. per doz., delivered in United States; $1.75 per 100, buyer paying transit.

Admiral de Constantinople. Large red flowers, tipped orange.

Monstre Rouge. Very handsome, large, deep crimson scarlet.

Monstre Cramoisie. Splendid deep crimson, extra large flowers, with large black star-shaped centre.

Markgraff van Baden. Yellow, striped with scarlet and green.

Lutea Major. Large, bright yellow.

Perfecta. Yellow, striped red.

PRICES.—Any of the above-named Parrots, 3 for 12c., 40c. per doz., delivered in United States; or $2.50 per 100, buyer paying transit.

The Collection of above 6 named sorts—1 bulb each for 25c., 3 bulbs each, 70c., 6 bulbs each, $1.25, 12 bulbs each, $2.00, delivered in United States.

A FEW GEMS IN LATE OR MAY-FLOWERING TULIPS.

Bouton d'Or. The richest and most beautiful pure golden yellow of all Tulips. Handsome globe-shaped flowers. 3 for 10c., 30c. per doz., delivered in United States; $2.00 per 100, buyer paying transit.

Breeder Tulips. (Mother Tulips.) Late-flowering Tulips, with immense flowers of solid colors, including scarlet, crimson, violet, etc. Choice mixed, 3 for 10c., 30c. per doz., delivered in United States; $1.75 per 100, buyer paying transit.

Bridesmaid. Brilliant scarlet, striped pure white, very distinct and beautiful. 3 for 10c., 35c. per doz., delivered in United States; $2.25 per 100, buyer paying transit.

May Blossom. Pure white, slightly striped and variegated with red. A grand flower of fine shape. 5c. each, 50c. per doz., delivered in United States; $3.50 per 100, buyer paying transit.

Elegans. A grand Tulip, very showy, rich crimson scarlet, large flower, pointed petals. 3 for 10c., 35c. per doz., delivered in U. S.; $2.00 per 100, buyer paying transit.

White Swan. A grand pure white late-flowering Tulip; it is the only white that we know of in late Tulips, and is just what is wanted for contrasting with the reds, yellows and purples, which heretofore predominated in late Tulips. "White Swan" grows about 12 inches high, a strong healthy grower with broad leaves; the flowers are extra large, broad petaled, and of pure silky white. 3 for 10c., 30c. per doz., delivered in United States; $2.00 per 100, buyer paying carriage.

True Tall Scarlet Gesneriana. (Hortensis.) (See cut.) Flowers of enormous size, on strong stems, the most durable of all Tulips, as it holds its beautiful color and keeps perfect several weeks. Color, rich crimson scarlet with glittering blue-black centre. 3 for 10c., 35c. per doz., delivered in United States; $1.75 per 100, buyer paying transit.

... NEW "DARWIN" TULIPS ...

Darwin Tulips belong to the late or May-flowering section. The flowers are very large, of symmetrical form, and are borne on tall, strong stems, often two feet high. They by far surpass in colors and brilliancy anything before known in Tulips. The colors are so glowing and bright that in the sunlight the effect is fairly dazzling. They include almost every conceivable color and shade, from the daintiest blue to the darkest violet, from soft rose to the most brilliant red, and from light brown to what is believed to be the darkest black in the floral world. The magnificent appearance of the beds of Darwin Tulips as we saw them in the introducer's grounds in Holland, in the full flush of their beauty, defies description.

PRICES.—Darwin Tulips, mixed colors, 5c. each, 60c. per doz., delivered in United States; $2.50 per 100, buyer paying transit.

NAMED VARIETIES OF DARWIN TULIPS.

Anton de Bary. Rich purplish wine.
Buys Ballot. Light carmine red.
Coros. Fiery rich blood scarlet.
Decamps. Mahogany red.
Europe. Bright orange crimson.
La Petit Blondin. Silvery lilac, shaded white.

Mad. Lethierry. Rosy flesh color.
Mad. Bosboom Toussaint. Dark cherry crimson.
Prof. Balfour. Dark brownish blood red.
Prof. McOwan. Violet blue.
Reve de Jeunesse. Lavender.
Terpsichore. Heliotrope, shaded claret.

Price for any of the above named Darwin Tulips, 5c. each, 60c. per doz., delivered in United States; $3.50 per 100, buyer paying transit.

The Collection of above 12 named Darwins, 1 bulb each, 60c., 3 bulbs each, $1.50, 6 bulbs each, $2.75, delivered in the United States.

DARWIN TULIP.

DO NOT FORGET to avail yourself of the Premiums offered on 2d page of cover.

DOUBLE TULIPS have massive flowers of brilliant and varied colors, shades and markings, and being double the flowers last much longer in bloom than single varieties, and in consequence, when singles and doubles are planted in conjunction the "time of the Tulips" is greatly prolonged. Double Tulips are beautifully adapted for beds on the lawn, in the garden, and for mingling in clumps of half a dozen or more around the edges of shrubbery. They are robust growers and exceedingly effective. The early sorts (those indicated by an A or B) do splendidly when grown in pots for winter blooming, but must be forced much slower than Single Early Tulips, by keeping them cooler.

The letters following the varieties indicate their earliness; those marked "A" flower together, and those marked "B" follow, etc. The figures indicate the average height in inches.

Tulip Bulbs are delivered free in the United States at the single and dozen price, but at the 100 and 1,000 price purchaser pays transit.

Agnes. *B7.* Bright fiery scarlet, large double flower. 3 for 12c., 40c. per doz., $2.25 per 100.

Alba Maxima, *B8.* A fine new double white. 3 for 12c., 45c. per doz., $2.50 per 100.

Arabella. *(Double Prosorpine.) A9.* Carmine rose flowers. 4 inches across, extra fine. 3 for 12c., 45c. per doz., $2.50 per 100.

Count Leicester. *B7.* Orange and yellow feathered. 3 for 12c., 40c. per doz., $2.25 per 100.

Crown of Gold. *(Couronne d'Or.) B10.* Large flower, very double, rich golden yellow, shaded orange. 6c. each, 60c. per doz., $4.00 per 100.

Crown of Roses. *(Couronne des Roses.) B9.* A magnificent Tulip, large, very double, of a rich, dark satiny, rose carmine color, shaded cerise, fine bedder. 6c. each, 60c. per doz., $4.25 per 100.

Duc Van Thol, double, red and yellow. *A7.* Very early, dwarf and fine for forcing. 2 for 5c., 28c. per doz., $1.25 per 100.

Duc Van Thol, double, rosy carmine. *A7.* 3 for 10c., 30c. per doz., $1.50 per 100.

Duc Van Thol, double, scarlet. *A8.* Crimson scarlet, good color. 3 for 10c., 35c. per doz., $2.00 per 100.

Duke of York. *B10.* Lovely variety, very double, carmine rose, edged broadly with white, suffused rose, showy and fine. 3 for 10c., 30c. per doz., $1.75 per 100.

Gloria Solis. *A9.* A grand flower, deep crimson, with broad golden margin, very large. 3 for 10c., 30c. per doz., $1.75 per 100. *(See cut.)*

Grand Alexander. *B6.* Dwarf in growth, but a large, very double, showy flower, mahogany red, broadly edged deep golden yellow. 3 for 10c., 35c. per doz., $2.00 per 100.

Helianthus. *A10.* Large, showy flower, deep crimson, edged deep golden yellow. 3 for 10c., 30c. per doz., $1.75 per 100.

Imperator Rubrorum. *A9.* Splendid bright scarlet, yellow base, a fine full double. 3 for 12c., 45c. per doz., $2.50 per 100.

Le Blason. *A9.* A beautiful variety, of delicate rose, shaded and striped with white, extra fine. 5c. each, 50c. per doz., $4.00 per 100.

La Candeur. *B8.* Pure white, very full and large, extensively grown for bedding. 3 for 10c., 30c. per doz., $1.50 per 100.

La Citadelle. *B9.* Purplish red, bordered yellow, fine and large. 3 for 10c., 35c. per doz., $2.00 per 100.

Lady Fitzharding. Cherry rose and white, beautiful, good for forcing. 3 for 10c., 35c. per doz., $2.00 per 100.

Lady Grandison. *A7.* Dwarf, vermilion scarlet, extra fine, good bedder. 3 for 10c., 35c. per doz., $1.75 per 100.

Leonardo-da-Vinci. *A9.* An extra fine, large Tulip, orange scarlet, edged with yellow, very brilliant. 3 for 12c., 45c. per doz., $2.50 per 100.

Lion d'Orange. *A8.* Brilliant orange, dwarf, extra choice, good forcer. 3 for 12c., 40c. per doz., $2.25 per 100.

Murillo (Albino.) *B8.* Magnificent blush white, shaded rose, large flowers, fine for forcing. 5c. each, 60c. per doz., $3.00 per 100.

Double Purple Crown. *A9.* Dark crimson maroon, velvety and fine. 3 for 10c., 30c. per doz., $1.50 per 100.

Princess Alexandra. *B7.* Dwarf, with a very large double flower, reddish brown, edged yellow. 3 for 10c., 30c. per doz., $1.75 per 100.

Double Queen Victoria. *B8.* Rich, glowing black scarlet. 3 for 10c., 35c. per doz., $2.00 per 100.

Rex Rubrorum. *B9.* Bright crimson scarlet, superb for bedding, large and showy. 5c. each, 50c. per doz., $2.75 per 100.

Rose Blanche. *B9.* Extra large, pure white, slightly tipped green. Good form. 3 for 10c., 35c. per doz., $2.00 per 100.

Rosine. *A13.* Blush white, tinged rose. 3 for 12c., 40c. per doz., $2.25 per 100.

GLORIA SOLIS.

6 bulbs of one variety sold at dozen rates; 25 at 100 rates. 1,000 rates on application.

DOUBLE EARLY TULIPS—Continued.

Tulip bulbs are delivered free in the United States at prices here quoted by the single and dozen, but at the 100 and 1,000 price purchaser pays transit.

Rubra Maxima. *.18.* Dark, dazzling carmine scarlet, lower half of outer petals flamed with green. 3 for 10c., 35c. per doz., $2.00 per 100.

Salvator Rosa. *.17.* Beautiful deep rose, flamed with white, fine for forcing. 6c. each, 60c. per doz., $4.00 per 100.

Titian. *.17.* Claret red, petals tipped white, yellow base. 3 for 10c., 30c. per doz., $1.75 per 100.

Tournesol. *.19.* Splendid showy Tulip, orange scarlet, with broad yellow tips and yellow base. Very large double flower. 3 for 12c., 40c. per doz., $2.25 per 100.

Tournesol Yellow. *.19.* Bright golden yellow, shaded orange. Very fine, large and showy flower, forces well. 6c. each, 60c. per doz., $4.00 per 100.

Velvet Gem. *.16.* Deep red mahogany, margined with yellow. Flowers large and beautiful. 3 for 10c., 35c. per doz., $2.00 per 100.

Virgilius. *.19.* An exquisite, richly colored Tulip, deep carmine, fading to blush at the edges. Large, full double flower. 5c. each, 50c. per doz., $2.75 per 100.

MIXED DOUBLE EARLY TULIPS.

Double Early Mixed Tulips. First quality, 20c. per doz., delivered in the United States, or, buyer paying transit, for $1.50 per 100, $6.00 per 1,000.

Henderson's Special Mixture of Double Early Tulips. Made up from Named Sorts with proper proportions of bright colors—all of uniform heights and blooming together. 25c. per doz., delivered in the United States, or, buyer paying transit, for $1.00 per 100, $8.50 per 1,000.

Entire Collection of Named Double Early Tulips as offered on pages 22 and 23.

35 bulbs, 1 each of 35 varieties, all named, $1.25, delivered in the U. S.
105 " 3 " 35 " " 3.50, " "
210 " 6 " 35 " " 6.00, " "

COLLECTIONS OF EARLY DOUBLE TULIPS FOR BEDS.

Bed "M."

BED OF DOUBLE EARLY TULIPS, "M."

This beautiful bed is circular, 7 feet across or 21 feet in circumference, requiring a total of 230 bulbs planted 5 inches apart. **Price** of either of the below combinations, $5.00, purchaser paying transit.

Yellow, White and Scarlet Combination, for Bed "M."

Centre Cross requires.......100 Count Leicester..............Yellow.
Four Corners require........ 40 La Candeur..................White.
Two Outside Rows require... 90 Rex Rubrorum................Scarlet.

Bed "N."

Bed "N" of Double Early Tulips.

This bed is 18 feet in circumference or 6 feet across, requiring a total of 200 double tulips planted 5 inches apart.

Price of either of below combinations, $4.50, buyer paying transit.

Scarlet and White Combination, for bed "N." No. 1 requires 100 double scarlet Tulips (Agnes); Nos. 2, 3 and 4, 33 each, white (Alba Maxima).

Crimson, Yellow, White and Pink Combination, for Bed "N." No. 1 requires 100 double crimson Tulips (Rubra Maxima); No. 2, 33 white (La Candeur); No. 3, 33 pink (Duke of York); No. 4, 33 Tournesol Yellow.

Crimson, Pink and Orange Combination, for Bed "M."

Centre Cross requires........100 Purple Crown...............Crimson.
Four Corners require....... 40 Salvator Rosa..............Pink.
Two Outside Rows require... 90 Lion d Orange.............Orange.

DOUBLE LATE FLOWERING TULIPS.

These commence blooming in this latitude about the first week in May, and are very showy companions to the Bizarre, Byblœmen and Parrot Tulips.

Admiral Kingsbergen. (12 in.) Mahogany red with golden yellow base, gradually verging into the red toward the ends of the petals. A very large double flower. 3 for 10c., 30c. per doz., $1.75 per 100.

Blanche borde pourpre. (12 in.) Wine red, bordered white. 3 for 10c., 30c. per doz., $1.50 per 100.

Bleu Celeste. (14 in.) Violet blue, very distinct. 3 for 10c., 30c. per doz., $1.50 per 100.

Bonaparte. (15 in.) Brownish red. 3 for 10c., 30c. per doz., $1.50 per 100.

Gloria Mundi. (16 in.) Lavender ground striped with white and claret, very large double flower. 3 for 10c., 35c. per doz., $2.00 per 100.

La Belle Alliance. (14 in.) Large violet red, striped with white. 3 for 10c. per doz., $1.75 per 100.

Louis d'Or. (14 in.) Sturdy stems holding flowers upright, very double, golden yellow, interior slightly splashed red. 3 for 12c., 40c. per doz., $2.25 per 100.

Marriage de ma Fille. (14 in.) Magnificent large flower, white, feathered with crimson, one of the finest late doubles grown. 5c. each, 50c. per doz., $3.50 per 100.

Pæony Gold. (14 in.) Golden yellow with red stripes and featherings. 3 for 10c., 30c. per doz., $1.75 per 100.

Pæony Red. (14 in.) Deep crimson red. 3 for 10c., 30c. per doz., $1.50 per 100.

Yellow Rose. (12 in.) Large, very double golden yellow flowers, fine bedder and sweet-scented. 3 for 10c., 30c. per doz., $1.50 per 100.

Mixed Double Late Flowering Tulips. 25c. per doz., delivered in U. S., or buyer paying transit, for $1.00 per 100, $8.00 per 1,000.

Entire Collection of Double Late Flowering Tulips.

11 bulbs, 1 each of the above 11 sorts, for.......................................$0.35
33 " 3 " " " 11 .... 1.00 } Delivered free in the United States.
66 " 6 " " " 11 .... 1.75

DO NOT FORGET TO AVAIL YOURSELF OF THE PREMIUMS ALLOWED ON ORDERS. SEE 2d PAGE OF COVER.

NARCISSUS
EMPRESS.

From a photograph of one bulb bearing fourteen flowers.

THE wonderful creations in the way of new varieties have awakened an interest and enthusiasm among the lovers of flowers that has placed this, "**The Flower of the Poets**," in the front rank of popularity, and they merit all the praise that can be bestowed upon them. Appearing, as they do, just after bleak winter, they turn our gardens, lawns and woodland walks into gorgeous masses of gold and silver, with a fragrance that is enchanting. They are equally valuable for growing in pots for winter flowering and are prized for pot culture, flowering in the house with the Hyacinth, and should be cultivated in the same way, 3 roots at least in a 4 or 5 inch pot. The cut flowers of Daffodils are much in demand for bouquets and vases, and some of the varieties are forced in immense quantities by florists for cut flowers in winter. Narcissus are of the easiest culture, and all, excepting the Polyanthus varieties on page 28, are perfectly hardy, *though the latter live through the severe winters—if well protected—and produce larger and more abundant flowers than when grown in the greenhouse.* Our collection has been much enriched, and embraces, in addition to the old favorites, many magnificent new hybrids.

NOTE—In describing Narcissus we have referred to the "perianth" and "trumpet." The latter is the long centre, funnel-like tube, and the "perianth" is the surrounding row of petals or leaves.

LARGE TRUMPET NARCISSUS.
ALL YELLOW VARIETIES.

Narcissus bulbs at the single and dozen price are delivered free in the United States, but at the 100 and 1,000 rate purchaser pays transit.

Golden Spur. One of the grandest Daffodils, with extra large, bold, rich yellow flowers, foliage very broad and striking. It is of unusually robust habit, and unsurpassed for gardens, pots or forcing. Early and extremely free-flowering. (*See cut.*) 10c. each, $1.00 per doz., $6.00 per 100.

Maximus or Kale's Beaten Gold. One of the largest and most beautiful of this class. Very large flower of rich, deep golden yellow. Remarkable for an elegantly twisted perianth. Trumpet deep golden, large, well flanged and deeply fluted. The darkest and richest yellow of all Daffodils. (*See cut on opposite page.*) 10c. each, $1.00 per doz., $6.00 per 100.

Henry Irving. A magnificent flower, with beautifully frilled trumpet, nearly two inches wide at the mouth, of rich golden yellow color. The petals of the perianth are very wide, overlapping, color bright yellow. A most perfect flower, received a first-class certificate. 10c. each, $1.00 per doz., $6.00 per 100.

Obvallarius. The famous "Tenby Daffodil." Elegantly formed medium-sized flower of rich yellow. One of the earliest and excellent for forcing. One of the most perfect flowers in this family, and we recommend it highly for any purpose. 5c. each, 50c. per doz., $3.00 per 100.

Rugilobus. A very free bloomer—broad petaled flower 3 inches across, of pale yellow, large trumpet of golden yellow, a beautiful variety. 5c. each, 50c. per doz., $3.00 per 100.

Trumpet Major. Flower large and almost of a uniform deep golden. Highly prized as an early forcing sort and largely planted for outside bedding. 3 for 12c., 40c. per doz., $2.25 per 100.

Emperor. One of the finest Daffodils in cultivation, entire flower of the richest yellow, trumpet of immense size, and the petals of the perianth are so broad they overlap and measure 3½ inches across. Grand for cutting. 15c. each, $1.50 per doz.

Countess of Annesley. A very vigorous and fine large-flowering variety, perianth light canary yellow; large, well-expanded trumpet of rich golden yellow. 15c. each, $1.50 per doz.

NARCISSUS GOLDEN SPUR.

COLLECTIONS "ALL YELLOW" LARGE TRUMPET NARCISSUS.

8 Bulbs, 1 each of the above 8 sorts for $0.85		delivered in		
24 " 3 " " " 8 " " 1.75		United		
48 " 6 " " " 8 " " 3.25		States.		

6 sold at dozen rates, 25 at 100 rates.   1,000 rates on application.   Delivered free in the U. S., except where noted.

LARGE TRUMPET NARCISSUS.

TWO-COLORED (White and Yellow) VARIETIES.

Narcissus Bulbs at the single and dozen price are delivered free in the United States, but at the 100 and 1,000 rate buyer pays transit.

Empress. A magnificent, large variety, bold and erect. A rare beauty. Some give this the palm of being the best of the two-colored trumpets. Perianth white and of great substance, trumpet rich yellow. 15c. each, $1.50 per doz., $10.00 per 100. (*See cut on preceding page.*)

Grandee. Flowers large, with broad white perianth, the petals of which are wide and overlapping, tube a magnificent yellow, beautifully imbricated; a flower of great substance, and grand for cutting. 15c. each, $1.50 per doz., $10.00 per 100.

Horsefieldi. The "Queen of the Daffodils." Very large flowers of pure white, with rich yellow trumpet. Very stately and beautiful, and one of the most elegant for ladies' corsage wear. The flowers are the largest of this class, very early and free-blooming, a grand cut flower, and eagerly sought after as one of the finest. (*See cut.*) 12c. each, $1.25 per doz., $8.00 per 100.

Princeps. Very early. This is one of the most popular winter-flowering and forcing varieties grown. Flowers large, perianth sulphur, with an immense rich yellow trumpet. 5c. each, 50c. per doz., $3.00 per 100.

Scoticus. The "Scotch Garland Flower," or "Large-flowering Lent Lily." Large flowers of creamy white, bright yellow trumpet, elegantly flanged and serrated. 3 for 12c., 45c. per doz., $2.50 per 100.

Michael Foster. Broad petaled, creamy white perianth, with large bright yellow trumpet, a dwarf bushy growing variety, very free-blooming. 10c. each, $1.00 per doz., $5.50 per 100.

COLLECTIONS OF "TWO-COLORED" LARGE TRUMPET NARCISSUS.

6 bulbs, 1 each of above varieties,	$0.50	} Delivered
18 " 3 " " "	1.75	in
36 " 6 " " "	3.00	} U. S.

NARCISSUS HORSEFIELDI.

MIXED TRUMPET NARCISSUS.

Plant liberally of these; they are cheap, and will make a grand show in the garden border, and are unequaled for cutting. Price, 25c. per doz., $1.25 per 100, $9.00 per 1,000.

MALE'S "BEATEN GOLD" OR MAXIMUS.

ALBICANS.

All=White Trumpet Narcissus

Albicans. The "White Spanish Daffodil." Creamy white, with a silvery white trumpet, slightly flushed with primrose and elegantly recurved at the brim. (*See cut.*) 10c. each, $1.00 per doz., $5.50 per 100.

Tortuosus. Very robust, perianth twisted, trumpet creamy white, passing to snow white, delightfully perfumed. 15c. each, $1.50 per doz., $10.00 per 100.

Pallidus Præcox. The "White Pyrenees Daffodil." One of the earliest and most beautiful varieties; both trumpet and perianth are clear sulphur-white; one of the best for forcing. The flowers, though not extra large, are very freely produced. 7c. each, 75c. per doz., $5.00 per 100.

COLLECTIONS OF ALL-WHITE LARGE TRUMPET NARCISSUS.

3 bulbs, 1 each of above sorts, for	$0.30	} Delivered
9 " 3 " " "	.80	in
18 " 6 " " "	1.50	} U. S.

DO NOT FORGET TO AVAIL YOURSELF OF YOUR PREMIUM. SEE SECOND PAGE OF COVER.

MEDIUM TRUMPET NARCISSUS, OR EUCHARIS-FLOWERED ...... DAFFODILS. ......

NARCISSUS BULBS we deliver free in the United States at the single and dozen price, but at the 100 and 1,000 price purchaser pays transit.

Medium Trumpet Narcissus are among the most useful of

all the Daffodil family for cutting, for bouquet, corsage and decorative purposes. The flowers are very large, averaging from 3 to 5 inches across with only medium sized trumpets, and on that account are preferred by many for their chaste and graceful form. They are all perfectly hardy and of equal value for open ground planting, flowering in the spring, or for growing in the house for winter flowers.

SIR WATKIN

ALL YELLOW VARIETIES.

Barri Conspicua. The finest of this class, long-stemmed flowers with beautiful broad petaled perianth 3½ inches across, sparkling canary yellow with deep golden cup richly edged with orange scarlet; awarded a first-class certificate. 10c. each, $1.00 per doz., $6.00 per 100.

Cynosure. Large flowers, 3½ inches across, light yellow, with rich deep yellow broad cup; a great beauty, and one of the best for either forcing or open ground planting. 3 for 12c., 40c. per doz., $2.25 per 100.

Figaro. A grand variety, with large long-stemmed flowers 3½ inches across, perianth lemon yellow, the petals of which are very broad. Cup deep rich golden yellow, edged with reddish orange. 3 for 12c., 40c. per doz., $2.25 per 100.

Sir Watkin. "The Giant Chalice Flower," or "Big Welshman." A gigantic variety, immense long-stemmed flowers, sometimes measuring 5½ inches across, being the largest variety grown. Color a rich light yellow, with a very large dark cup, tinted with orange. Awarded a first-class certificate by the R. H. S. (*See cut.*) 15c. each, $1.50 per doz., $9.00 per 100.

*D*O not forget to avail yourself of our Premium Offer. See Second Page Cover.

ALL WHITE VARIETIES.

Circe. (Duchess de Brabant.) Fine pure white; the cup is slightly tinted with light straw at first, but changes to white. A very distinct and pretty variety. 3 for 12c., 40c. per doz., $2.25 per 100

Leedsii. (Type.) Pure white star-like flowers. Cup at first slightly tinted, but changes to white. A very pretty variety, exceedingly sweet-scented, a free bloomer, and fine for forcing as well as for open ground planting. (*See cut.*) 3 for 10c., 35c. per doz., $2.00 per 100.

LEEDSII

WM. WILKS

TWO-COLORED (White and Yellow) VARIETIES.

Backhousi. Wm. Wilks. Large broad petaled flower, sulphury white, with orange yellow cup; very distinct and beautiful. (*See cut.*) 10c. each, $1.00 per doz., $6.00 per 100.

Lorenzo. A beautiful variety, with creamy white perianth, the petals of which are long and the whole flower large, cup golden yellow. 5c. each, 50c. per doz., $2.75 per 100.

Nelsoni Major. An excellent second early variety, flowers very large, perianth 3 inches across, of creamy white, broad and fluted cups ½ inch long of deep golden, grand for cutting. 10c. each, $1.00 per doz., $6.00 per 100.

Stella. One of the first in flower and wonderfully free-blooming. White, star-shaped flowers 3 inches across, with bright yellow cups, very beautiful and one of the most popular forcing and cutting varieties for the London flower market. 3 for 10c., 35c. per doz., $2.00 per 100.

TYPES OF MEDIUM TRUMPET NARCISSUS.

MIXED MEDIUM TRUMPET NARCISSUS.

If you wish to grow a lot of beautiful, graceful, fragrant flowers for cutting purposes, either inside or in the open garden, or if you only want a grand show in your garden border, plant liberally of these mixed Narcissus. Price 25c. per doz., delivered free in the United States, or $1.25 per 100, $9.00 per 1,000 bulbs, buyer paying transit.

Collections of MEDIUM TRUMPET NARCISSUS.

Including the All Yellows, All Whites, and the Two-colored Classes offered on this page.

10 bulbs, 1 each of all sorts offered on this page, for	$.65	Delivered in		
30 " 3 " " " " "	1.75	United		
60 " 6 " " " " "	3.25	States.		

6 bulbs of one variety supplied at dozen rates, 25 at 100 rates, 250 at 1,000 rates. All bulbs are delivered free in the U. S., except where noted.

NARCISSUS POETICUS ORNATUS.

Narcissus bulbs are delivered free in the United States at the single and dozen price, but at the 100 and 1,000 price purchaser pays transit.

Narcissus Poeticus, with Miniature Saucer=shaped Cups.

Poeticus Biflorus. Twin flowers, borne on one stalk, pure white with yellow cup. 3 for 10c., 30c. per doz., $1.50 per 100.

Poeticus Burbidgei. An early-flowering Poeticus, fully a month earlier. Flowers 2½ inches across, round, broad, overlapping petals. Pure white, cup edged with orange-scarlet, very fine for both forcing and outside planting. 3 for 10c., 30c. per doz., $1.50 per 100.

Poeticus Ornatus. The improved "Poeticus." A grand flower for cutting purposes. Larger and more symmetrical than the old variety and considerably earlier. Beautiful pure white flowers, with saffron cup, tinged with rosy scarlet. A magnificent cut flower. (*See cut.*) 3 for 10c., 35c. per doz., $2.00 per 100.

Poeticus. The "Pleasant's Eye," or "Poet's Narcissus," pure white flowers, with orange cup, edged with crimson. Splendid cut flower. 2 for 5c., 20c. per doz., 75c. per 100, $6.00 per 1,000.

Poeticus Vanessa. Perianth yellow, shaded to primrose, cup yellow. A perfect flower. 5c. each, 50c. per doz., $3.00 per 100.

COLLECTIONS OF POETICUS NARCISSUS.

15 bulbs,	3 bulbs each,	5 varieties	...............$0.50 } Delivered in
30 "	6 "	5 "	.... .85 } United States.
60 "	12 "	5 "	.... 1.50 }

HOOP PETTICOAT NARCISSUS.

Bold and shapely flowers. They are gems for pot culture and bear from 6 to 12 flowers to each bulb. For planting in groups around the edges of lawns, and especially for edgings for beds, this class is superbly adapted.

Bulbocodium. The "Large Yellow Hoop Petticoat," rich golden yellow. 10c. each, $1.00 per doz., $6.00 per 100.

Citrinus. The "Large Sulphur Hoop Petticoat," large sulphur flowers, unique and beautiful. 5c. each, 50c. per doz., $3.50 per 100.

"Algerian White Hoop Petticoat." Pure snow-white, very early, will bloom at Christmas if potted in September. 5c. each, 50c. per doz., $3.50 per 100.

COLLECTIONS OF HOOP PETTICOAT NARCISSUS.

3 bulbs each	....$0.50
6 "	.... .85
12 "	.... 1.75
	Delivered in the United States.

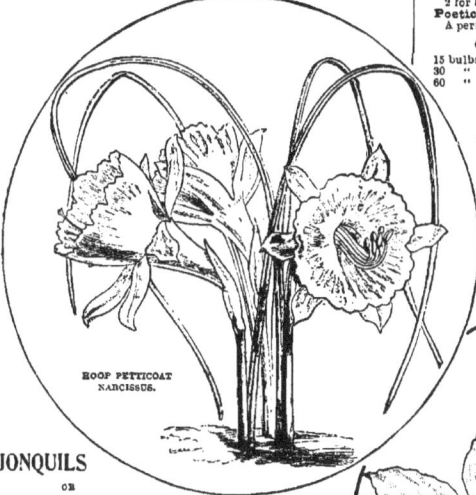

HOOP PETTICOAT NARCISSUS.

JONQUILS

OR

NARCISSUS JONQUILLA.

Jonquil bulbs are delivered in the United States free at the single and dozen price, but at the 100 and 1,000 price purchaser pays transit.

Much prized for their charming golden and deliciously sweet-scented flowers, perfectly hardy, and flowering very early in the spring; they are also admirably adapted for winter forcing. (*See cut.*)

Single Jonquil. The well-known favorite, delicately scented and beautiful for forcing. Rich yellow, very fragrant. 2 for 5c., 20c. per doz., 75c. per 100, $6.00 per 1,000.

Double Jonquil. Heads of small but very double deep golden yellow flowers, powerfully scented and good for forcing. 5c. each, 50c. per doz., $4.50 per 100.

Campernelle. Large, 6-lobed yellow flowers, 4 to 6 on a stem, fine for forcing, and the one generally preferred by florists. (*See cut.*) 2 for 5c., 25c. per doz., $1.25 per 100.

Rugulosus. (The Giant Jonquil.) Broadly imbricated perianth with large wrinkled cup. Full yellow. 3 for 10c., 30c. per doz., $1.75 per 100.

COLLECTIONS OF JONQUILS.

12 bulbs,	3 each of above 4 varieties, for	...............$0.35 } Delivered in
24 "	6 " " 4 "	.... .65 } United States.
48 "	12 " " 4 "	.... 1.25 }

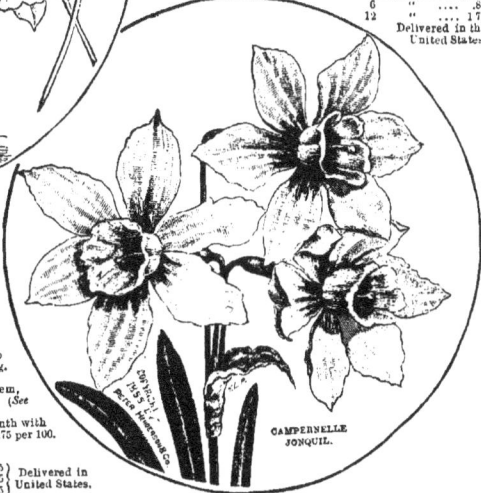

CAMPERNELLE JONQUIL.

POLYANTHUS

Narcissus Bulbs are delivered free in the United States at the single and dozen price, but at the 100 and 1,000 price purchaser pays transit.

THE *Polyanthus* varieties of Narcissus are not only beautiful but deliciously sweet-scented, and of the easiest culture; very free - flowering, and suitable for window garden, conservatory or garden, continuing long in bloom. They bear tall spikes of bloom, bearing from six to twenty - four flowers each. The pure white petals and gold cups of some varieties, the yellow with deep orange cups of others, and the self whites and yellows render them great favorites.

NARCISSUS.

YELLOW VARIETIES.

Grand Soleil d'Or. Rich yellow, with reddish orange cup — a favorite bedding and cutting variety. 6c. each, 60c. per doz., $4.00 per 100.

Sunset. A remarkably beautiful and distinct variety. The flowers are borne in clusters of four or more, and are of the most exquisite fragrance; the petals are canary yellow, and the cup rich orange-scarlet. Price, 6c. each, 60c. per doz., $4.25 per 100.

Newton. Fine yellow, with orange cup, very free bloomer. 6c. each, 60c. per doz., $4.00 per 100.

TWO - COLORED VARIETIES.

(Yellow and White.)

States General. Clusters of white flowers with citron - colored cups shading to white. Very early. 6c. each, 60c. per doz., $3.50 per 100.

Gloriosus. Immense trusses of pure white, with primrose - colored cups. Very early, splendid to force. (*See cut.*) 5c. each, 50c. per doz., $3.25 per 100.

Grand Monarque. Large white flowers, with lemon yellow cup. Can be successfully grown in water like the "Chinese Sacred" mentioned below. 6c. each, 60c. per doz., $4.25 per 100.

Double Roman. Clusters of white flowers with double citron-colored cups. It is very early, and is grown largely for forcing. 5c. each, 50c. per doz., $3.00 per 100.

Mixed Polyanthus Narcissus. 30c. per doz., $1.25 per 100, $9.00 per 1,000.

WHITE VARIETIES.

Bouquet Sans Pariel. Large trusses of pure white flowers. 6c. each, 60c. per doz., $4.25 per 100.

White Pearl, True. (Louis Le Graud.) Large, pure satiny white flowers, exquisite. 7c. each, 75c. per doz., $5.00 per 100.

Paper White. Pure snow-white flowers in clusters. This variety is perhaps more largely forced for cut flowers than any other, millions of them being used. 5c. each, 50c. per doz., $2.75 per 100.

Paper White Large-flowered. This new variety is of vigorous growth and early bloom, with immense individual flowers and larger truss. (*See cut.*) 6c. each, 60c. per doz., $4.00 per 100.

—THE—

Chinese Sacred or Oriental Narcissus.

The "Shui Sin Far," or Water Fairy Flower, Joss Flower, or Flower of the Gods, etc., as it is called by the Celestials, is a variety of Narcissus, bearing in lavish profusion chaste flowers of silvery white, with golden yellow cups. They are of exquisite beauty and entrancing perfume. It is grown by the Chinese, according to their ancient customs, to herald the advent of their new year and as a symbol of good luck.

The bulbs are grown by a method known only to themselves, whereby they attain great size and vitality, ensuring luxuriant growth and immense spikes of flowers; in fact, the incredibly short time required to bring bulbs into bloom (four to six weeks after planting) is one of the wonders of nature. "You can almost see them grow," succeeding almost everywhere and with everybody. They do well in pots of earth, but are more novel and beautiful grown in shallow bowls of water, with enough of fancy pebbles to prevent them from toppling over when in bloom. A dozen bulbs started at intervals will give a succession of flowers throughout the winter. (*See cut.*) Price (*True Chinese Grown Bulbs, extra large*), 12c. each, $1.25 per doz., $8.00 per 100.

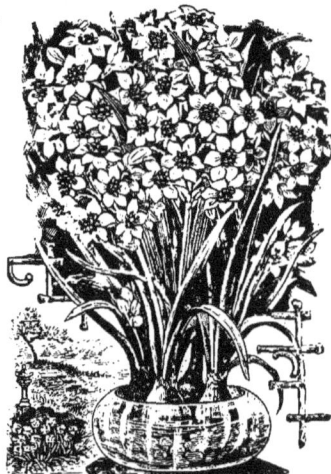

THE CHINESE SACRED NARCISSUS.

---

6 Bulbs of one variety sold at dozen rates, 25 at 100 rates.   Delivered free in the U. S., except where noted.

DOUBLE ...
NARCISSUS or
... Daffodils.

.. Narcissus Bulbs we deliver free in the United States at the single and dozen price, but at the 100 and 1,000 price buyer pays transit.

VON SION.

ORANGE PHŒNIX NARCISSUS.

ALBA PLENA ODORATA. "The Double White Poet's Narcissus or Gardenia Flowered Daffodil." Double snow-white Gardenia-like flowers, exquisitely scented. 3 for 10c., 30c. per doz., $1.50 per 100.

INCOMPARABLE FL. PL. "Butter and Eggs." Full double flowers of rich yellow with orange nectary. Splendid variety for either forcing for winter-cut flowers or for open ground planting. 3 for 10c., 30c. per doz., $1.75 per 100.

ORANGE PHŒNIX. "Eggs and Bacon." Beautiful double white flowers with orange nectary. Splendid for pot culture and cutting or garden decoration. (See cut.) 5c. each, 50c. per doz., $3.50 per 100.

SULPHUR (OR SILVER) PHŒNIX. "Codlins and Cream." Large creamy white flowers with sulphur nectary. Exquisite corsage flower, and fine for growing in pots. 12c. each, $1.25 per doz., $6.00 per 100.

VON SION. (Telamonius Plenus.) The famous "Old Double Yellow Daffodil." Rich golden yellow perianth and trumpet. One of the best forcing sorts, immense quantities being grown for this purpose in Europe and America. (See cut.) 5c. each, 50c. per doz., $3.00 per 100.

Mixed Double Narcissus. 25c. per doz., $1.25 per 100, $9.00 per 1,000.

Collections of Double Narcissus ...

15 bulbs, 3 each of the above 5 named varieties.,	$0.75	Delivered free
30 " 6 " " "	5 "	"	1.35	in the
60 " 12 " " "	5 "	"	2.50	United States.

Our Mixed Narcissus AT REDUCED PRICES...

These are suitable for growing in masses for garden decoration, and are grand for cutting. They are perfectly hardy, and will flourish and increase. The mixtures include many beautiful varieties.

	Per doz.	Per 100.	Per 1,000.
Mixed Large Trumpet Narcissus	$0.25	$1.25	$9.00
" Medium "	.25	1.25	9.00
" Double Narcissus	.25	1.25	9.00
" Polyanthus	.30	1.25	9.00

Narcissus Bulbs are delivered free in the United States at the single and dozen price, but at the 100 and 1,000 price buyer pays transit.

Do not forget to avail yourself of our Premium offer on second .... page of cover. ....

MIXED NARCISSUS.

Crocus ..

"The Harbingers of Spring."

THE CROCUS is one of the earliest flowers of spring, and occupies a prominent place in every garden. When planted as an edging in triple lines of one or more colors, the effect is striking. No spring display surpasses that of Crocus; the broad wavy bands of golden-yellow, striped, purple, or of pure white, when they expand their blossoms in February and March, are incomparable. In lawns and pleasure parks, planted in the grass, the Crocus is extremely effective. For several years past Crocus, Snowdrops, Daffodils, etc., have been planted in our parks and lawns, in wild places, woodland walks, etc., to the great delight of visitors and the enhanced decoration of the parks. In one of the parks thousands of Crocus were used in the grass, and the effect was matchless.

Crocus Bulbs are delivered free in the United States at dozen and 100 price, but at the 1,000 price purchaser pays transit.

Crocus in Mixed Colors.

	Per doz.	Per 100	Per 1,000
Blue and purple, mixed	$0.10	$0.50	$4.00
Variegated and striped, mixed	.10	.50	4.00
White, mixed	.10	.50	4.00
Large yellow	.10	.50	4.00
All colors, mixed	.10	.45	3.50

AUTUMN FLOWERING CROCUS, PARKINSONI.

Named Large=Flowered Crocus.

LARGE NAMED YELLOW CROCUS.

Yellow Mammoth. Large bulbs, each producing a number of large golden yellow flowers, 15c. per doz., $1.00 per 100, $8.00 per 1,000.

Yellow First Size. Bright yellow. 10c. per doz., 50c. per 100, $4.00 per 1,000.

Cloth of Gold. Golden yellow, striped brown, small flowers, but very profuse. 10c. per doz., 50c. per 100, $4.00 per 1,000.

LARGE NAMED WHITE CROCUS.

15c. per doz., 75c. per 100, $6.00 per 1,000.

Caroline Chisholm. Pure white.

Princess of Wales. Splendid, large, pure white.

Mont Blanc. Large and fine white.

Theba. The purest of whites, very free-blooming.

LARGE NAMED STRIPED & VARIEGATED CROCUS.

15c. per doz., 75c. per 100, $6.00 per 1,000.

Albion. Violet, striped lavender and white, extra large flower.

Cloth of Silver. Silvery white, striped lilac, small flowers, but very profuse.

Imperati. Flowers in January, bluish-white, purple stripe, yellow throat, sweetly scented.

La Majesteuse. Immense white flowers, striped lilac.

Sir Walter Scott. White striped, purple, extra.

Ne Plus Ultra. Purple, variegated with white.

LARGE NAMED BLUE AND PURPLE CROCUS.

15c. per doz., 75c. per 100, $6.00 per 1,000.

Baron Von Drunow. Large, purple.

David Rizzio. Large, dark purple.

Garibaldi. Extra large, dark purple.

Lilaceus. Light blue, extra.

Autumn Flowering Crocus.

Colchicum Autumnale. (Meadow Saffron.) Very effective and handsome hardy plants, the flowers of which come through the ground without the leaves in the fall, the leaves appearing the following spring. The flowers comprise many shades of white, purple, rose, striped, etc. They make lovely borders or margins to beds. Mixed sorts, 5c. each, 60c. per doz., delivered in United States free, or $3.50 per 100, buyer paying transit.

Parkinsoni. A beautiful rare variety from the Greek Archipelago; large flowers of white, tessellated and barred with rose and purple, like a checkerboard; very odd. (*See cut.*) 10c. each, $1.00 per doz.

6 sold at dozen rates, 25 at 100 rates, 250 at 1,000 rates. Delivered free in United States, except where noted.

MISCELLANEOUS BULBS.

For Autumn Planting.

All bulbs are delivered free in the United States, except where noted.

ACHIMENES.

These are splendid and profuse summer-blooming plants for the conservatory or window decoration; flowers of many charming colors ranging through all shades from white to crimson, including many that are beautifully spotted.

Mixed Varieties. *(Ready in November.)* 10c. each, $1.00 per doz.

AGAPANTHUS UMBELLATUS.

The Great African Lily.

These are noble ornaments in pots or tubs for lawns, terraces or piazzas or for the decoration of the greenhouse. Foliage luxuriant and graceful; flowers bright blue, borne in clusters of 20 to 30, and measure fully a foot across. The flower stalks frequently attain a height of 3 feet, the flowers opening in succession for a long period during the summer and autumn. *(See cut.)* Ready in November. 20c. each, $2.00 per doz.

ALSTROMERIA.

Peruvian Lilies. Tuberous-rooted plants, robust and abundant blooming, with beautiful, large lily-like flowers of great beauty, borne in clusters during the summer; colors —crimson, rose, yellow, purple, etc., shaded and marked. They are splendid for cutting, being of much substance and lasting in perfection for a long time. Splendid subjects for either pot culture or for planting out in frames. 2 to 4 feet. *(See cut.)*

Mixed Varieties. *(Ready in November.)* 10c. each, $1.00 per doz.

ALLIUM.

The varieties we offer are among the most beautiful for pot culture or garden decoration. They are of the easiest culture.

Azureum. Truly beautiful, either for pot or garden culture, being quite hardy, flowers deep azure blue, borne in large umbels; height, 1 to 2 feet. 15c. each, $1.50 per doz.

Aureum (Moly or Golden Allium). One of our most showy border plants, perfectly hardy, bearing large golden yellow flowers in June. A very old favorite, and fine for naturalizing in the garden, where it forms large clumps; height, 1 foot. 3 for 10c., 30c. per doz., $1.75 per 100.

Hermitti Grandiflorum. A splendid winter-flowering and forcing variety; the flowers last a long time after being cut. The flowers are large, of immaculate whiteness, and continue to bloom from December to the end of April. *(See cut.)* 3 for 10c., 30c. per doz., $1.75 per 100.

Neapolitanum. Another excellent variety for winter flowering now extensively forced by florists for cut flowers, being of pure white with green stamens, borne in large loose umbels; height, 15 to 18 inches. 3 for 10c., 30c. per doz., $1.75 per 100.

Ostrowskianum. A beautiful new species from Asia Minor, with large heads of beautiful rose-colored flowers on stalks 2 feet high; very early, free flowering and hardy. 15c. each, $1.25 per doz.

6 bulbs sold at dozen rates, 25 at 100 rates; delivered in United States free, except where noted.

ALSTROMERIAS.

AGAPANTHUS.

ALLIUM GRANDIFLORUM.

AMARYLLIS FORMOSISSIMA.

...Amaryllis...

Johnsonii. "The Barbadoes Spice Lily." Congenial conditions—climate and soil—produce a strong growing variety of Amaryllis, popularly known in the West India Isles as "The Spice Lily." Probably the most magnificent and gorgeous flowering bulbous plants known. Their immense flowers, richness of coloring and regal habit are simply incomparable. The bulbs are very large, 8 to 12 inches in circumference, and produce, with great certainty, extra strong spikes of bloom, from 18 inches to 3 feet high, bearing enormous trumpet shaped flowers of great substance, averaging 6 to 10 inches across, of rich and glowing scarlet, intensified by a contrasting wide white stripe through each petal. For pot culture in the window, conservatory or greenhouse they are well adapted, and when in bloom in the winter and spring months no flower can approach their beauty—spikes frequently produce 4 to 8 flowers each, and often 2 and 3 spikes are produced at one time. Extra large bulbs (*now ready*), 30c. each, $3.00 per doz.

Amaryllis Vittata Hybrida. These are unnamed seedling hybrids, and will produce some new varieties of exceptional beauty. The colorings and markings are exquisite, the bulbs are very large, 12 inches and over in circumference, and are of sufficient strength and age to produce magnificent flowers during the winter or spring. (*Ready in November.*) Price, $1.00 each, $10.00 per doz.

Aulica Platypetala. "Lily of the Palace." Flowers large and extremely handsome, glowing crimson, tipped green. Splendid winter bloomer. $1.25 each, $12.00 per doz.

Formosissima—(Jacobœan Lily) (*Now ready*). A quaintly shaped beautiful flower of grand dark scarlet, free blooming; forces well and can be grown in water like Hyacinths; if the bulbs are kept dry during winter, they can be planted in the open ground in the spring, and will flower during the summer. *See cut.* 15c. each, $1.75 per doz.

AMARYLLIS DEFIANCE.

Defiance. This is a gem, being a sturdy grower, flowering repeatedly during the season. Flowers extraordinarily large, of velvety carmine red, striped and flecked and suffused with white. $1.25 each, $12.00 per doz.

Equestris. Bright light scarlet, with a white star-like throat, running out into bars in the centre of the petals, a very free bloomer. 25c. each, $2.50 per doz.

Amaryllis Belladonna Major (Belladonna Lily), An autumn blooming variety of extreme beauty and fragrance. The spikes grow from 2 to 3 feet high, each carrying from six to a dozen beautiful flowers "Sweet as Lilies," of silvery white, flushed and tipped with rose. (*See cut.*) 15c. each, $1.75 per doz.

For Amaryllis Sarniensis see "Nerine"; for A. Lutea see "Sternbergia"; for A. Purpurea see "Vallota."

DO NOT FORGET THE PREMIUM YOU ARE ENTITLED TO.

♣♣ See 2d page Cover. ♣♣

AMARYLLIS BELLADONNA MAJOR.

ANEMONES.

HIGHLY ornamental spring and summer flowering plants, having both single and double flowers, the colors of which are very beautiful. Anemones are also splendid for pot culture, for flowering in the house or conservatory during winter.

Double Poppy Flowered. (*Coronaria fl. pl.*) The flowers of this class are very double—and are surrounded at the base with large guard petals—for cutting they are grand, and we know of no class of plants with more gorgeous flowers, two or three colors usually being blended in the flowers of each variety. Mixed colors. 3 for 12c., 30c. per doz., $2.00 per 100.

Single Poppy Flowered. (*Coronaria.*) Large, beautiful, saucer-shaped, poppy-like blossoms, during mild seasons, or in sheltered situations, they flower continuously throughout the winter, spring and early summer. (*See cut.*) Mixed colors. 3 for 10c., 25c. per doz., $1.50 per 100.

Fulgens. Rich dazzling scarlet flowers. It is invaluable for cutting, as it lasts a long time in water; it is adapted for pot culture—flowering during the winter in the house—and is perfectly hardy for garden work—flowering in the spring. 3 for 12c., 35c. per doz., $2.75 per 100.

Fulgens, fl. pl. A double-flowering variety of the above, very beautiful. (*See cut.*) 6c. each, 50c. per doz., $3.50 per 100.

Blanda. The earliest and largest-flowered of the spring-blooming Anemones, blooming with Snowdrop and Crocus, colors range from pure white to deep blue. The flowers are 1¼ inches across. It is a lovely variety and as hardy as a rock. It spreads itself in large clumps, grows freely. If potted at intervals from August to September, and housed afterwards in the greenhouse, they may be had in flower from November to February. 3 for 10c., 30c. per doz., $2.00 per 100.

ANEMONE FULGENS, FL. PL.

HARDY... ANEMONES.

These are beautiful for permanent situations, where they soon form large clumps of great beauty when in bloom; a situation, partially shaded, suits them to perfection.

Hepatica Angulosa. (*Ready in November.*) One of the finest spring blooming varieties; it is a vigorous grower and blooms profusely large sky-blue flowers, on stalks 6 to 9 inches high. 20c. each, $2.00 per doz.

Apennina Mixed Blue, White and **Rose** Varieties. (*Ready in November.*) Beautiful flowers, as large as a fifty cent piece, elegantly cut foliage, they bloom profusely in early spring, also beautiful grown in pots. ½ foot. 3 for 10c., 30c. per doz., $2.00 per 100.

SINGLE ANEMONE.

ANOMATHECA.

...ARUM DRACUNCULUS...

Flowers one foot long, purple, red and black-blue; stem beautifully marbled, leaves handsome, resembling a small palm; a rapid grower, making a curious and ornamental pot plant. (*See cut.*) 15c. each, $1.50 per doz.

ARUM DRACUNCULUS.

...ANTHOLYZA...

Very showy and stately, having the appearance of Gladiolus. They are hardy south of Washington, and can be grown in cold sections in a cold frame, or they may be planted out in the spring, flowering from July to September. For grouping in beds or shrubberies, their brilliant, long tubular flowers of scarlet, black, green, etc., in happy combinations, and tall spikes of bloom render them very effective; also fine for cut flowers. **Mixed Varieties.** 3 for 12c., 35c. per doz., $2.50 per 100.

...ANOMATHECA CRUENTA...

A charming plant for either pot culture or for blooming out of doors, hardy south of Washington, but requiring the protection of a cold frame in cold climates; their dwarf stature, brilliant and profuse bloom continued for a long period, render them very popular; flowers bright rosy carmine. Height, ¾ foot. (*See cut.*) 3 for 12c., 35c. per doz., $2.50 per 100.

"Henderson's Bulb Culture." Price, postpaid, 25c.; or given free as a Premium. See Second page of Cover.

BRODIÆA.

⇥ BABIANA. ⇤

A charming genus, with leaves of the darkest green, thickly covered with downy hairs, and hearing showy spikes of flowers, characterized by their rich self-colors, or the striking contrast of very distinct hues in the same flower. They vary in color from the richest carmine to the brightest lilac, many of them being sweet-scented. As they are not hardy north of Washington, they should have the protection of a cold frame. They are very successfully grown in pots. Five or six bulbs in a five-inch pot make lovely and useful specimens. Height, 6 to 9 inches.

Mixed Varieties. 5c. each, 50c. per doz., $3.50 per 100.

BRODIÆA.

Showy, half-hardy California bulbs, with lovely umbels or clusters of red, blue or white tubular flowers, borne on stems one to two feet high. They are easily forced, and may be grown in the greenhouse or cold frame, or if planted out in spring in clumps or masses they flower very freely in June or July.

B. Capitata. Early; large heads, color lavender. 5c. each, 50c. per doz.
B. Coccinea. "The Floral Fire-cracker Plant." Crimson, tipped green. (*See cut.*) 5c. each, 50c. per doz.
B. Lactea. White flowers banded green. 5c. each, 50c. per doz.
B. Volubilis. (Twining Hyacinth.) A unique novelty; delicate rose-pink. When in bud, the stems commence to twine and often reach 5 feet by the time the flower unfolds. 15c. each, $1.50 per doz.
B. Mixed Varieties. Containing many beautiful colors. 3 for 10c., 35c. per doz., $2.75 per 100.

SPOTTED LEAF CALLA.

LITTLE GEM CALLA.

BLACK CALLA.

❧ CALLAS. ❧

Calla Æthiopica, or Lily of the Nile. This old favorite White Calla Lily is too well known to require any description. We offer dry roots, as they are superior for forcing and winter-flowering purposes; they come into bloom quickly and require less room, making less foliage. Extra large dormant roots, 25c. each, $2.50 per doz. First size, 15c. each, $1.50 per doz.

Little Gem Calla Lily. This little pigmy rarely exceeds 12 inches in height, and blooms most abundantly. The flowers are more than half the size of the common variety, and therefore can be used with telling effect in bouquets. It is in every way superior as a house plant to the larger-growing variety. 20c. each, $2.00 per doz. (*See cut.*)

Spotted Leaf Calla. (*Richardia Alba Maculata.*) This plant is always ornamental, even when not in flower, the dark green leaves being beautifully spotted with white; in other respects the plant is the same as the old favorite White Calla, excepting being of smaller habit. In addition to its usefulness as a pot plant it makes a fine thing for planting in the garden in the summer, being very effective. (*See cut.*) 20c. each, $2.00 per doz.

Yellow Calla. (*Richardia Hastata.*) This is identical in all respects to the well-known White Calla, excepting that the flowers are of light yellow. Choice and rare. $1.00 each, $10.00 per doz.

Black Calla. (*Arum Sanctum.*) A magnificent and remarkable variety from the Holy Land. The plant produces one large flower the shape of a Calla, but from 14 to 18 inches long and 4 inches broad, and of a rich, dark purple color and green underneath, somewhat wavy at the borders and curled at the smaller end. The spathe rising from the centre of the flower is about 10 inches long, velvet-like and quite black. It is raised on a slender but vigorous stalk of brown-red, shading to green at the upper end. The leaves are large and very wavy, of a rich green color, veined light green, and resemble exactly those of the Calla Æthiopica. The whole plant makes a most stately and elegant appearance. (*See cut.*) **Price,** dry bulbs as collected, 15c. each, $1.50 per doz.; large cultivated bulbs, 25c. each, $2.50 per doz.

Red Calla. See Novelties, page 8. 25c. each, $2.50 doz.

COLLECTION OF CALLAS. One bulb each of the above 6 varieties, including the Red Calla, largest sizes of each, $1.75, delivered free in the United States.

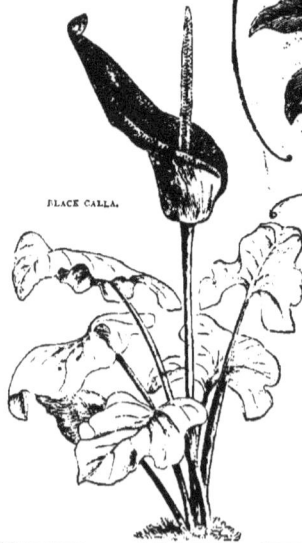

6 of one variety sold at dozen rates, 25 at 100 rates, delivered free in the United States, except where noted.

.. Chionodoxa, or Glory of the Snow ..

These are praised by all as the most exquisite of spring-flowering plants, and when grown in quantities under shrubs the effect is strikingly beautiful. They produce flower spikes bearing ten to fifteen lovely Scilla-like flowers. They are perfectly hardy, and may be planted as an edging to a bed or in clumps or masses, where they are doubly welcome, flowering early in the season with the Snowdrops, and lasting a long time in perfection. They will thrive well in any good garden soil, and are admirable for pot culture for winter blooming in the house and for forcing for cut flowers. (*See cut.*)

Lucilliæ. Charming bright blue, with large, clear white centre. Select bulbs. 3 for 10c., 25c. per doz., $1.25 per 100, $10.00 per 1,000.

Sardensis. Intense deep true Gentian blue; in masses its brilliant color catches the eye at a great distance. The flowers, though smaller than those of other Chionodoxas, are more numerous. Select bulbs. 3 for 10c., 25c. per doz., $1.25 per 100, $10.00 per 1,000.

Gigantea. A great acquisition, differing from all others of this family by its unusually large flowers of lovely lilac blue, with conspicuous white centre. Is thoroughly hardy, and a perfect gem for spring decoration in masses in the garden, and when grown in pots for winter flowering it is beautifully effective. (*See cut.*) Select bulbs. 3 for 10c., 30c. per doz., $1.75 per 100, $15.00 per 1,000.

Tmolusi. Very large flowers of dark indigo blue. Select bulbs. 3 for 10c., 30c. per doz., $1.75 per 100.

Calochortus

These are the *Butterfly Tulips* or *Mariposa Lilies* of California, and possess such delicacy and brilliancy of color that the most unobservant are struck with their characteristic beauty. The flowers somewhat resemble a tulip in shape; are of many brilliant colors, ranging through various shades of white, lilac, blue, crimson, yellow, etc., some being wonderfully spotted, veined, edged or tipped with gold. Planted in May in the open border they flower in June and July. They also succeed admirably when grown in a cold frame, and form very handsome specimens if planted in a five-inch pot for winter flowering. (*See cut.*)

Mixed Varieties. Containing many beautiful sorts. 3 for 10c., 30c. per doz., $2.25 per 100.

CHIONODOXA, OR GLORY OF THE SNOW.

Camassia

The "Quamash" of the Indians. Perfectly hardy, thriving in sheltered and partially shady situations; very handsome and valuable for flower borders; the stout flower stalks grow from 2 to 3 feet high and bear 20 or more large blue flowers, each 2 inches across. A large clump in bloom is very effective. The flowers are fine for cutting, lasting for a long time in water. 3 for 10c., 30c. per doz., $2.00 per 100.

Crown Imperials

Well-known spring-blooming, stately, hardy border plants, with clusters of immense pendent bell-shaped flowers surmounted with a tuft of green leaves. They are very effective, and if left undisturbed for years they form gigantic and picturesque groups of gorgeous colors. The variegated-leaved ones are especially excellent, and also very effective for conservatory decoration among dwarf-growing plants. (*See cut.*)

Aurora. Red. 15c. each, $1.75 per doz.

Crown upon Crown. Several whorls of flowers one above the other. 20c. each, $2.00 per doz.

Sulphurea. Sulphur yellow, slightly striped red. 25c. each, $2.50 per doz.

Golden Striped. (*Folia Aurea Striata.*) Rich green, striped with golden yellow; the bright red flowers are strikingly effective against the beautiful variegated foliage. This variety is of much value for pot culture and forcing. (*See cut.*) 25c. each, $2.50 per doz.

Mixed Varieties. 15c. each, $1.50 per doz.

CROWN IMPERIAL. (GOLDEN STRIPED.)

Six bulbs of one variety sold at dozen rates, 25 at 100 rates. Delivered free in United States, except where noted.

❧ CYCLAMEN. ❧

These are among the most beautiful and interesting winter and spring flowering bulbs for the window and greenhouse. Not only are the flowers of striking beauty, but the foliage is highly ornamental; consequently they are very decorative, even when not in bloom. There are no plants better adapted for pot culture, and few that produce such a profusion of bloom; the flowers range through many shades, pink, crimson, white, etc., some being beautifully spotted. Most of them are, moreover, delicately fragrant. (See cut.)

Persicum, Mixed. Dry bulbs. 15c. each, $1.50 per doz.

❧ GIANT CYCLAMEN. ❧ ❧

The flowers of this magnificent strain are of extraordinary size and of great substance. The leaves are proportionately large and beautifully marked. (See cut.)

Giganteum, Rose Color.
     "   **Crimson.**   } Dry bulbs.
     "   **White.**

Price for any of above, 30c. each, $3.00 per doz., or the set of 3 varieties, 75c.

Giganteum, Mixed Varieties. Dry bulbs. 25c. each, $2.50 per doz.

YELLOW... "DAY LILY."

(Hemerocallis Flava.)

Very ornamental hardy plants, having elegant foliage and handsome flowers of bright yellow, delicately perfumed. They are of the easiest culture in any ordinary garden soil, and form admirable clumps. The flowers, of bright light yellow, are very fragrant and fine for cutting; they are somewhat ephemeral, but are produced successively and in great abundance. Although perfectly hardy, they bear forcing well in a temperature of 50 degrees. 2 to 3 feet high. 20c. each, $2.00 per doz. (Ready in October.)

For the new "Japanese Day Lilies," see Novelties, page 10.

GIANT PERSIAN CYCLAMEN.

ERYTHRONIUM.

DICENTRA CUCULLARIA.

DICENTRA, OR ❧ BLEEDING HEART.

(Ready in November.)

Spectabilis (*Bleeding Heart*). One of the most ornamental of spring-flowering plants, with elegant green foliage and long drooping racemes of pink and white heart-shaped flowers. This is deemed one of the finest of all hardy garden plants, and is frequently forced for greenhouse or conservatory decoration. 20c. each, $2.25 per doz.

Cucullaria (*Dutchman's Breeches*). Similar to the above, one of our most beautiful spring flowers; fern-like foliage and a profusion of white blossoms. (See cut.) 20c. each, $2.25 per doz.

ERYTHRONIUM GRANDIFLORUM.

Giant Dog's-Tooth Violet.

A beautiful large flowering variety, perfectly hardy; the foliage is charmingly variegated, and a mass of 15 or 20 plants is a pretty sight, even when not in flower, but when the graceful flowers are in bloom the effect is matchless. The plants luxuriate in rather moist, partially shady positions, and do very nicely when grown in pots in frames and brought into the conservatory or window garden for winter blooming. This variety bears from 6 to 12 yellow or cream-colored flowers on stems 12 to 18 inches high. (See cut.) 3 for 12c., 35c. per doz., $2.50 per 100.

❧ ERYTHRONIUM JOHNSONII. ❧

See Novelties, page 9.

FREESIA.

Refracta Alba. This is one of the most popular and charming bulbs for pot culture, flowering in the winter and spring in the conservatory or window garden. 6 or 8 bulbs should be planted in a 4-inch pot. They force readily and can be had in bloom by Christmas, if desired, and by having a dozen or more pots started in the cold frame they can be brought in at intervals, thereby keeping up a continuous display of bloom throughout the winter; the flowers are produced 6 to 8 on stems about 9 inches high, and are particularly useful for cutting, remaining in good condition, kept in water, for two weeks; the flowers are a pure white with a yellow blotched throat, and are exquisitely fragrant. *(See cut.)* Extra large bulbs, 3 for 10c., 25c. per doz., $1.25 per 100, $10.00 per 1,000.

Leichtlinii Major. The new yellow Freesia. *(See Novelties, page 8.)*

FRITILLARIA.

A group of dwarf spring-flowering plants, bearing singular large pendent bell-shaped flowers of white, purple, bronze, black or yellow, most of which are striped, splashed or checkered in the most fantastic fashion; they are invaluable for pot culture, and exceedingly pretty when grown in large clumps in the border in a dry situation. *(See cut.)*

Meleagris. (Guinea Hen Flower.) Bell-shaped flowers of various colors, yellow, white, black, purple, striped and splashed, and checkered in the most curious way. Mixed varieties, 3 for 10c., 30c. per doz., $1.00 per 100.

Meleagris Latifolius. A larger-growing variety of the above, with broad foliage and large checkered flowers, earlier and a decided improvement. 5c. each, 50c. per doz., $3.00 per 100.

Aurea. A lovely new species, as rare as it is beautiful, bearing large golden yellow bell-shaped flowers, which are curiously checkered with black-brown spots. It is perfectly hardy, beginning to flower in March, continuing till May. It is admirably adapted for planting out-of-doors in partial shade. May also be grown in pots for greenhouse or winter decoration, where its curious flowers prove a constant source of admiration. 10c. each, $1.00 per doz., $6.00 per 100.

Kamtchatkensis. Also known as the "Siberian Black Lily." The stems attain a height of from 12 to 18 inches, bearing pendulous bell-shaped flowers of richest purple-black color, very hardy, flowering in spring. 20c. each, $2.00 per doz.

FRITILLARIAS.

FREESIA.

GLADIOLUS.

Early-flowering Hardy Varieties.

These beautiful early-flowering Gladiolus are greatly prized on account of their blooming in June and July, if planted in the autumn in dry soil and protected with a covering of about 6 inches of straw, leaves or litter. If planted in cold frames they will flower as early as May, and this perhaps is the better way in very cold localities. These Gladiolus are also invaluable for flowering in the greenhouse in pots for winter bloom. The colors and markings are very handsome; the bulbs can be kept dormant until spring, and then planted in open ground if preferred. *(Brody in November.)*

The Bride. *(Colvilli Alba.)* Very beautiful; purest white. 3 for 10c., 25c. per doz., $1.75 per 100.

Mixed Early-flowering Hardy Gladiolus. 3 for 10c., 25c. per doz., $1.75 per 100.

GLOXINIA.

GLOXINIAS.

SOME GRAND ERECT FLOWERING VARIETIES.

(READY IN NOVEMBER.)

Gloxinias, as well known, are the most beautiful flowering plants that we have for the decoration of greenhouse and window during the spring and summer months. They are of easy culture and good bulbs produce from 50 to 100 flowers, 3 to 4 inches across, of the most exquisite and gorgeous colors, many of which are magnificently spotted, mottled and blended, during their season; as many as 20 are frequently open at one time, and the effect is charming. The following 5 varieties we selected from a large collection in Europe this season as being the most beautiful and distinct; they are all of the erect or "look you in the face" class, as gardeners call them, which are now considered so much superior to the pendulous varieties. *(See cut.)*

Defiance. Glowing crimson-scarlet, lustrous and rich.

Emperor Frederick. Bright rosy scarlet, bordered with a pure white band; strikingly beautiful.

Tiger Spotted. Beautiful spotted and mottled.

Kaiser Wilhelm. Velvety cerulean blue with deep white throat.

Mont Blanc. Pure satiny white; exquisite.

Price of any of the above-named varieties, 25c. each, $2.50 per doz., or the set of above 5 named varieties for $1.00.

Mixed Gloxinias. Large flowering, best varieties. 20c. each, $2.00 per doz.

6 bulbs of one variety supplied at dozen rates; 25 at 100 rates. All bulbs delivered free in the U. S., except where noted.

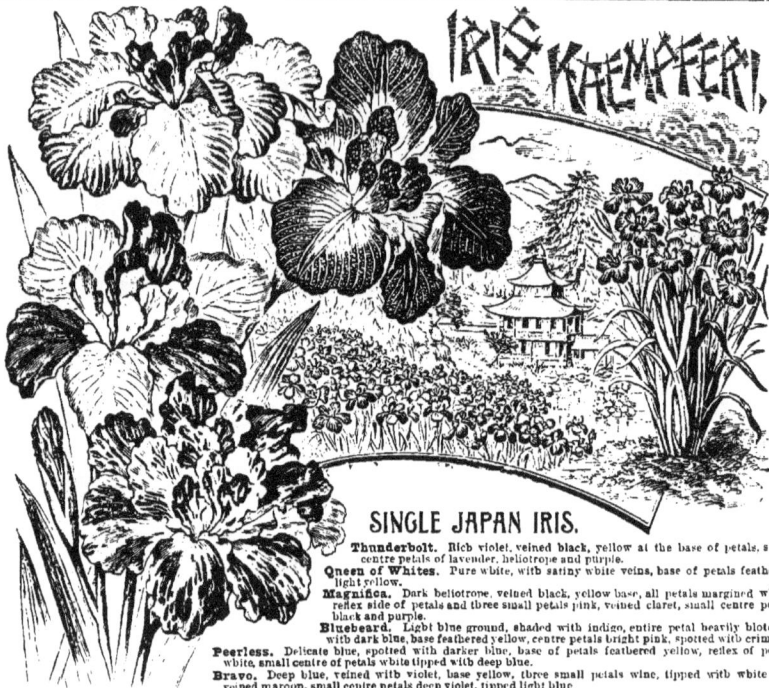

IRIS KAEMPFERI

Japan Iris.

(Ready In October.)

THE magnificence of these new Iris surpasses description. The flowers are of enormous size, averaging 8 to 10 inches across, and of indescribable and charming hues and colors, varying like watered silk in the sunlight, the prevailing colors being white, crimson, rose, lilac, lavender, violet and blue; each flower usually representing several shades, while a golden yellow blotch, surrounded by a halo of blue or violet, at the base of the petals, intensifies the wealth of coloring. The Japan Iris is perfectly hardy, and flowers in great profusion in July and August, and attains greatest perfection if grown in moist soil, or if plentifully supplied with water while growing and flowering. *(See cut.)*

MIXED JAPAN IRIS.

Single mixed, 20c. each, $2.00 dozen.
Double   20c. each, $2.00   "

(Delivered free in the U. S.)

SINGLE JAPAN IRIS.

Thunderbolt. Rich violet, veined black, yellow at the base of petals, small centre petals of lavender, heliotrope and purple.

Queen of Whites. Pure white, with satiny white veins, base of petals feathered light yellow.

Magnifica. Dark heliotrope, veined black, yellow base, all petals margined white, reflex side of petals and three small petals pink, veined claret, small centre petals black and purple.

Bluebeard. Light blue ground, shaded with indigo, entire petal heavily blotched with dark blue, base feathered yellow, centre petals bright pink, spotted with crimson, white, small centre of petals white tipped with deep blue.

Peerless. Delicate blue, spotted with darker blue, base of petals feathered yellow, reflex of petals white, small centre petals deep violet, tipped light blue.

Bravo. Deep blue, veined with violet, base yellow, three small petals wine, tipped with white and veined maroon, small centre petals deep violet, tipped light blue.

Speckled Beauty. White ground, minutely speckled with magenta and copper red, base of petals feathered with yellow, centre petals white, speckled with copper red.

Triumph. Ground color white, heavily blotched and spotted with garnet and light pink, base of petals yellow, small centre petals white, tipped with light claret.

Aurora. White shaded with rosy claret, dark picotee edged, entire flower veined white, base of petals feathered yellow, centre petals white, shaded light claret.

Curiosity. Pure white, shot with blue and violet, yellow feather in centre, centre petals white, tipped maroon and lavender.

Sensation. White ground, feathered yellow at base with surrounding zone of lavender, entire petal veined violet, inner petals light claret, veined maroon and shot white, centre petals bright blue, shot with white.

Gem. Lavender with margin of blue, yellow feather at base, reverse of petals lavender, entire flower veined and speckled with white, centre petals white, tipped dark violet.

PRICE for any of the above-named Single Japan Iris, 25c. each, $2.50 dozen.

DOUBLE JAPAN IRIS.

Snowball. Pure white, with satiny white veins and light yellow centre. Very double.

La Superb. Delicate blue, broadly margined with pure white, yellow centre, entire flower intricately veined with white.

Fascination. Bright blue, broadly margined with indigo, yellow centre. The three smaller inner petals pure white, tipped with claret, the entire flower veined white.

Conqueror. Deep carmine pink with band of white through centre of the petals, reverse of petals light pink, large veins of white through the entire flower. Very double.

Charmer. Pure white ground heavily veined with bright blue, base of petals ornamented with yellow feather, the three small inner petals bright blue and violet.

L'Unique. Ground color lavender, heavily veined with claret, base of petals feathered with yellow, small centre petals plum color.

Diversity. Rich deep blue, heavily veined with indigo, base of petals feathered with yellow, small centre petals creamy white, tipped with indigo, reverse of petals light blue.

Garnet. Rich garnet, lightly margined with white, dark maroon veins, base of petals feathered with yellow, reverse side of petals light rose, small centre petals violet and black.

Royal Purple. Rich violet, shaded deep purple, veined with black, base of petals feathered yellow, small centre petals varying from light lavender to dark purple.

Mikado. Rich crimson, with large veins of white, base of petals feathered yellow, small centre petals of lavender and plum color.

Perfection. Clear pink veined and speckled with dark pink, reverse side of petals bright red, centre yellow blotch, small centre petals creamy white, tipped with bright red.

Pride of Japan. Pure white ground, margined with pink and veined with satiny white, lightly speckled with pink, yellow feather at base, small centre petals white with same markings.

PRICE for any of the above-named Double Japan Iris, 25c. each, $2.50 dozen.

Either set of 12 single named or 12 double named, $2.25 ; or both sets for $4.00.

... GERMAN IRIS.
(Iris Germanica.)

This variety is the true "Fleur-de-Lis," the national flower of France. They are perfectly hardy, thrive anywhere, grow and bloom luxuriantly, particularly if plentifully supplied with water or if planted in moist situations, as on the banks of ponds, etc. Plants well established produce from 50 to 100 spikes of bloom, deliciously fragrant and fine for cutting. In beauty the flowers rival the finest Orchids, colors ranging through richest yellows, intense purples, delicate blues, soft mauves, beautiful claret reds, white, primroses and bronzes of every imaginable shade. *(See cut.)* Ready in November.

Fulda. Standards, soft lavender; falls, light and dark blue, veined and feathered.

Honorable. Standards, bright golden yellow; falls, finely feathered and veined yellow and maroon.

Queen of the Gypsies. Standards, smoked pearl and bronze; falls, feathered and veined with rich plum, white and buff.

Mad. Chereau. Standards, old gold beautifully frilled; falls, purple and white, fine.

Souvenir. Standards, bright yellow; falls, a veined network of yellow, buff and purple.

Stella. Fine creamy white.

Price for any of the above named sorts, 25c. each, $2.50 doz., or the set of six for $1.25.

Mixed German Iris. Containing many beautiful varieties. 20c. each, $2.00 per doz.

GERMAN IRIS.

SPANISH IRIS.

IRIS

Fœtidissima folia variegata. *(Variegated Gladwin.)* A very ornamental variety for pot culture and very desirable for open ground, of easy cultivation in almost any situation, but prefers a moist one; the flowers of bluish lilac are followed by large thrice-divided seed pods, showing the large, orange colored seeds, and are very ornamental; the foliage is also beautifully variegated with ivory-white, making the plant at all times very decorative. 25c. each, $2.50 per doz.

Histrio. Charming new species, bright blue, blotched yellow, and very early. 25c. each, $2.50 per doz.

Peacock. *(Pavonia.)* Pure white, with a bright blue spot on each petal. Fine for pots or garden culture. 1 foot. *(See cut.)* 10c. each, $1.00 per doz.

IRIS LORTETI.

IRIS

Reticulata. A lovely variety. Color, violet blue, lower petals of a darker shade, with gold and white stripes and veins, spotted with black. Very sweet-scented. 15c. each, $1.50 per doz.

Spanish. *(Hispanica.)* This type is well adapted for pot culture and forcing, blooming in the winter. They almost equal orchids in beauty of coloring and delicacy of perfume. The flowers are of great beauty, and cut in the bud state last from one to two weeks in bloom—longer than any other cut Iris. One root bears one stem bearing two flowers in succession a week apart. If planted in boxes and placed in cold frames until they show bud, then brought into heat, they can be had in bloom from March on, and will give great satisfaction.

British Queen. A massive flowering, pure white

Leonidas. Dark, violet blue.

Belle Chinoise. Large, deep golden yellow.

Olympia Creamy yellow and light blue.

Lilaceus. Grand flower of porcelain blue.

Prince of Orange. Orange yellow, bronze and porcelain blue; very large flower.

PRICE for any of the above named **Spanish Iris**, 5c. each, 50c. per doz., $3.00 per 100.

Mixed Spanish Iris, containing many beautiful varieties, 3 for 10c., 25c. per doz., $1.75 per 100.

English. *(Anglica.)* Large, handsome flowers, with rich purple, blue and lilac colors predominating. Grows 18 to 20 inches high. Perfectly hardy. Mixed varieties. 3 for 10c., 30c. per doz., $2.00 per 100.

Florentina. Very beautiful, pure violet slightly shaded with blue and with a yellow beard, deliciously violet-scented. The orris root of commerce is produced from this plant. 10c. each, $1.00 per doz.

Lorteti. It is considered one of the most beautiful Irises in the world. A native of Palestine. The flowers are as large as those of the remarkable "I. Susiana" and the coloring is very fine. The falls show a creamy or white ground, marked with small crimson-purple spots, and sometimes also veins, concentrated at the centre into a dark, crimson-purple "signal." The standards are nearly pure white and marked with thin violet lines. The often vivid yellow-crimson coloring of the styles gives, by reflex, a reddish shine to the standards. An exceedingly charming species; blooms in June. *(See cut.)* 50c. each, $5.00 per doz.

PEACOCK IRIS, OR IRIS PAVONIA.

Six bulbs of one variety sold at dozen rates, 25 at 100 rates. Delivered free in United States, except where noted.

IRIS—Continued.

Scorpion Iris. (*Alata.*) This is a gem. The plant only grows about a foot high. The flowers are very large, measuring four to six inches across, of a delicate lilac blue, with showy blotches of brightest yellow, spotted with a darker shade, the whole forming one of the richest combinations of color imaginable. One of the features of this rare plant is that its flowers are produced when our gardens are practically flowerless, commencing to bloom in October and producing a second crop of flowers in December if the weather be not too severe. It is of the easiest culture, but prefers a warm, dry sunny border. It is admirably adapted for pot culture for greenhouse or window garden. 10c. each, $1.00 per doz.

Susiana. (*The Mourning Iris.*) A remarkably handsome species with immense flowers, blush color, tinted with brown and covered with a network of dark lines; May-flowering; height, 1 foot. A remarkable variety. 50c. each, $5.00 per doz.

IXIAS.

The Ixia is a beautiful little winter-flowering bulb, with low, slender, graceful spikes of bloom. The colors are rich, varied and beautiful, the centre always differing in color from the other parts of the flower, so that the blossoms, expanding in the sun's rays, present a picture of gorgeous beauty. (*See cut.*)

Crateroides. Bright scarlet, the earliest of all, and grand for forcing. 3 for 10c., 30c. per doz., $2.00 per 100.

Wonder. A double variety. Very deep pink, sweetly perfumed. Extra. 5c. each, 50c. per doz., $4.00 per 100.

Mixed Varieties. Many beautiful colors. 2 for 5c., 20c. per doz., $1.25 per 100.

LACHENALIAS.

Beautiful early spring-flowering bulbs for conservatory and window-garden decoration. The spotted foliage and spikes of brilliant flowers render them exceptionally striking. They are very easily grown, and can be had in bloom by Christmas, if desired, and can be grown in cold frames if protected from frost. (*See cut.*)

Nelsoni. A new hybrid; without doubt the finest of the race, producing its large golden yellow flowers in long racemes with wonderful freedom, lasting in flower in a cool house nearly two months. Of the easiest culture; treated same as Hyacinths. 30c. each, $3.00 doz.

Pendula. Very strong-growing and handsome flowers, bright red, tipped with green and yellow. 25c. each, $2.50 doz.

Eubida. Deep red, freely spotted, one of the first in bloom. Very distinct and showy. 25c. each, $2.50 per doz.

IXIAS.

IXIOLIRION.

Tartaricum. An elegant and rare half-hardy bulbous plant of free growth, and producing grand spikes, two feet high, of splendid bell-shaped star-like flowers of rich purple, shaded with sky-blue. Bulbs planted out by us last autumn proved perfectly hardy with a slight protection, and bloomed beautifully in June; the bulbs can be kept dry and planted out in spring, or they may be grown in pots in cold frames and be brought into the conservatory toward spring for blooming. (*See cut.*) 15c. each, $1.50 per doz.

IXIOLIRION.

LENTEN OR CHRISTMAS ROSE.

(*Helleborus Niger.*)

A most beautiful class of hardy plants growing freely in almost any situation, flowering in great profusion in early spring, and if grown in the house, or in frames, will bloom from December all through the winter months. The flowers are two or three inches in diameter, pure waxy white (*See cut.*) (*Ready in November.*) 30c. each, $3.00 per doz.

Henderson's Bulb... Culture

Tells how to grow bulbs for winter flowering in the house or greenhouse, and for spring flowering in the garden. It treats on summer-flowering bulbs. It also tells how to "force" bulbs.

Price, postpaid, 25c., *or given free as premium on an order. See second page cover.*

LENTEN OR CHRISTMAS ROSE.

LACHENALIAS.

❧❧ LILY OF THE VALLEY. ❧❧

(Ready for delivery in November.)

The Lily of the Valley is one of the most useful and greatly admired plants grown; the modest bell-shaped flowers of purest white are highly prized for cutting purposes, and for flowering in pots in the winter they are exceedingly well adapted. Beautiful and most interesting ornamental designs for the parlor or conservatory may be produced by planting the Lily crowns in Crocus pots or in pyramidal pots made especially for this purpose, and pierced with holes. They will last several weeks in beauty. By taking them in at intervals, a succession of different designs may be kept up all winter. They are also forced in immense quantities by florists, but they are the most charming when grown in large patches, in partially shaded localities around the lawn, near the borders of streams, lakes, etc., being perfectly hardy. *(See cut.)*

Large-flowering German Single Crowns. 3 years old, for forcing, pot culture, or open ground planting. Per bundle of 25 crowns, 60c., $2.00 per 100; or, buyer paying transit, $15.00 per 1,000.

Henderson's Extra "Christmas Forcing" Single Crowns. The finest grade of Crowns in the world for early winter flowering, they will bear 12 to 16 large bells on strong stalks, *with foliage*, even when forced for extra early; the pips average large, plump and regular, with extra long roots. Per bundle of 25 crowns, 75c.; per 100, $2.50; or, buyer paying transit, $20.00 per 1,000.

Large Clumps of Lily of the Lily, for open ground planting, 30c. each, $3.25 per doz.; or, buyer paying transit, $20.00 per 100.

Fortin's Giant Lily of the Valley, for open ground planting. *See Novelties, page* 8.

HENDERSON'S "CHRISTMAS FORCING" LILY OF THE VALLEY.

LYCORIS, NERINE OR GUERNSEY LILIES.

Also known as "Japanese Hardy Amaryllis."

These bulbous plants are of great beauty; they belong to the Amaryllis family and are consequently adapted to pot culture in the greenhouse or window garden. Coming from colder climates than the tropical Amaryllis, they thrive under cooler treatment. They make splendid garden plants, and this is the popular way of growing them in Japan and China. Grown outside they flower towards autumn. L. Radiata, Aurea and Squamigera have proved to be hardy grown in a warm, sheltered sunny border, protected with a frame or mulching during the winter.

Aurea. Produces flower stems about 18 inches high, surmounted with from 12 to 18 deep yellow lily-like blossoms. 25c. each, $2.50 per doz.

Radiata, also known as **Nerine Japonica or Sarniensis** and the **"Guernsey Lily."** Flower stems about a foot long with clusters of from 8 to 12 fiery red lily-like flowers about two inches across; colors very brilliant, and in the sunlight glisten as if sprinkled with gold dust; the wavy, recurved petals and long red stamens give a graceful effect to the flower. These beautiful Lilies are perhaps the most popular and useful of the Amaryllis family; perhaps no bulbs bloom with more certainty and swiftness after potting; for autumn and winter flowering they are extremely beautiful. 10c. each, $1.00 per doz. *(See cut.)*

Squamigera. A very strong-growing and large-flowering variety, and perhaps the hardiest of the lot; it is frequently grown by the Chinese as a cemetery plant. The flower spikes are stout and solid, from 2 to 3 feet high, bearing umbels of from 5 to 7 large lily-like flowers about 4 inches in length, nearly twice the size of the other varieties. The color is a soft rosy pink, delicately tinged with silvery gray. 35c. each, $3.50 per doz.

Fothergilli Major. *See Novelties, page* 10. 60c. each, $6.00 per doz.

LYCORIS RADIATA, ALSO KNOWN AS NERINE JAPONICA OR SARNIENSIS.

THE BERMUDA
BUTTERCUP OXALIS.

THE BERMUDA "BUTTERGUP" ∴ OXALIS.

An Unrivaled Winter-flowering Pot Plant.

Of the Easiest Culture, Succeeding with Everybody.

THIS is one of the finest winter-flowering plants for pot culture that we have ever seen; it is such a strong, luxuriant grower that one bulb will be sufficient for a six or eight inch pot. Place in a dark, cool position for several days to root thoroughly, and remove to a sunny situation in the window or conservatory in a temperature of about 60° Fahr., and the great profusion of bloom produced in uninterrupted abundance for weeks will astonish and delight you. The flowers are of the purest bright buttercup yellow. Well-grown plants have produced as high as seventy flower stems at one time, and over 1,000 flowers in one season. The flowers, and frequently the leaves, fold up at night and open again the next morning, but when grown in a partially shaded situation the flowers remain open all the time. Properly treated, the plants will flower in six weeks from the time the bulbs are planted.

We do not claim the Bermuda Buttercup Oxalis to be strictly a new plant, but a greatly improved selection from Oxalis Cernua, grown in the congenial soil and climate of Bermuda until the bulbs have attained great strength, producing bulbs, plants and flowers larger and more luxuriant in all parts than the type. (*See cut.*) 10c. each, 3 for 25c., 75c. per doz., $5.00 per 100.

❀ OXALIS. ❀
Various Kinds.

Charming little half-trailing or bushy plants particularly adapted for pot culture and hanging baskets. The foliage alone is very attractive, and when in flower they are exceedingly pretty. The pots should be well filled, from six to a dozen bulbs in a five or six inch pot; for the smaller species three or four inch pots are large enough; they may be potted at any time during winter and placed near the glass or window to keep them stocky and dwarf.

Price for undermentioned named Oxalis, 2 for 5c., 25c. per doz., $1.75 per 100.

Boweii. Vivid rosy crimson, large.
Lutea. Splendid large canary yellow.
Lutea fl. pl. Very double bright yellow.
Mixed Oxalis. 2 for 5c., 20c. per doz., $1.50 per 100.

Alba. White.
Rosea. Rose.
Versicolor. Crimson and white.

❀ ORNITHOGALUM. ❀

Arabicum. (*Arabian Star of Bethlehem.*) A beautiful variety throwing up a tall spike bearing numerous large, milk-white, star-shaped flowers, with a black centre, and having a distinct aromatic perfume. They are decidedly pretty and interesting when grown in the garden, but are more largely grown for greenhouses and window decoration, being of the easiest culture. Largely forced by florists now for cut flowers. (*See cut.*) 6c. each, 65c. per doz., $4.00 per 100.

❀ PUSCHKINIA. ❀

Scillioides. A bulbous plant of great beauty, perfectly hardy, flower pearly white, richly striped with pale blue down the centre of each petal; its numerous flowers spring up from the centre of its peculiarly shaped leaves. It is admirably adapted for edgings and forming patches in front of mixed borders, flowering in April and May; height, four to eight inches. They do very nicely in pots for winter flowers, if started slow and cool. (*See cut.*) 3 for 10c., 35c. per doz., $2.50 per 100.

PUSCHKINIA.

ORNITHOGALUM ARABICUM

6 bulbs of one variety sold at dozen rates, 25 at 100 rates.
Delivered free in the United States, except where noted.

DOUBLE CHINESE... HERBACEOUS PÆONIES.

We herewith offer some of the newest and most beautiful varieties in cultivation. These noble plants are exceedingly effective; the profusion and duration of bloom, combined with handsome massive foliage, accommodating habit, and easy culture, render them one of the most popular hardy plants grown for lawn and garden decoration, or for mingling with shrubs or herbaceous plants in borders and wild gardens. The flowers are large, massive, perfect in outline and most beautiful (*Ready in October*)

DOUBLE CHINESE PÆONY.

Old Standard Varieties of ...PÆONIES...

Prices (*purchaser paying transit*).

	Each.	Doz.
Double White	$0 25	$2.50
Double Rose	.25	2 50
Double Crimson	.25	2 50
Double, Mixed Colors	.20	2.00

Postage extra, 5c. each, if wanted by mail.

...New Varieties of Double Pæonies...

Abel Carriere. Amaranth shaded with light violet; very double flower surrounded with large guard petals, extra fine.
Comte de Paris. Lovely rosy lilac with a centre of salmon, yellow and rose, very double, with large rose-colored petals in centre
Eugene Verdier. Rosy pink, a very large and brilliant flower, with blush centre; extra fine.
La Curiosite. Of a beautiful and brilliant rose color, centre of a soft rosy pink; a very fine flower of perfect habit; extra.
La Coquette. Lovely bright rose with chamois bands, centre rich brilliant red shaded with carmine; extra.
Mad. Lebon. Very large full double flower of the most brilliant cherry-red, with silvery reflections; superb.
Plenissima rosea superba. Large and splendid well formed flower of a lovely rose color, intermingled with salmon-colored petals; extra.
Pulcherrima. A beautiful and finely formed flower, outer petals yellow, centre crimson striped with carmine.
Queen Hortense. (*Hericartiana.*) Outer petals lovely violet rose, the centre rosy salmon; a very charming variety.
Reine des Francais. (*Umbellata.*) Salmon rose with a centre of white shaded with rose; a most beautiful and unique variety.
Triomphe du Nord. A lovely flower, rich violet rose shaded with the most delicate lilac; extra.
Victoire Lemoine. A large and beautiful variety, flowers well-formed, of a rich dark crimson amaranth.

Ready for delivery in October.

Price (*purchaser paying transit*), any of the above-named varieties, 50c. each; or your selection of 3 sorts for $1.25, 6 sorts, $2.25, or the entire set of 12 for $4.00. *Postage extra, 5c. each, if wanted by mail.*

CORAL PODDED PÆONY. (Pæony Corallina.)

A very beautiful single-flowering variety, with large rosy crimson blossoms in May; these are followed by very beautiful seed pods which open, displaying an interior of brilliant scarlet color, in which are thickly studded large glossy jet-black seeds; these remain on the plant for several weeks, making it an object of great beauty and attraction.

Price (*purchaser paying transit*), 60c. each, $6.00 per doz. If wanted by mail, add 6c. each for postage.

CORAL PODDED PÆONY.

FLUMED TREE PÆONY.

Plumed Japanese ... ... Tree Pæonias.

The flowers are all very double with the unique and beautiful addition of several of the centre petals projecting away beyond the circumference of the main flower—in the manner of a cluster of plumes.

"Silver Wedding." (*Fuji-arashi.*) Beautiful double pure white flowers, the centre petals projecting out like plumes.

"Heap of Treasure." (*Ko-jo-jima.*) Charming double flowers of deep rose edged light pink, centre projecting petals slightly lighter in coloring.

"Floral Mirror." (*Akebishishi.*) Fine double flowers of blush white, fading to pure white at edges of petals ; projecting centre petals of same color.

Price for any of the above 3 varieties, $1.50 each, or the set of 3 for $4.00. (*Can only be sent by express at purchaser's expense.*)

New Japanese Tree or ... Shrubby Pæonias.

(*Pæony Mouton.*)

The flowers are marvels in beauty, colorings and size, often 8 to 10 inches across, the petals just crinkled enough to give them, in the sunlight, the effect and sheen of crumpled satin. The plants are grafted on the roots of a strong growing Herbaceous Pæony, and in planting in the open ground, the plants should be put deep enough so the graft will be 3 or 4 inches below the surface of the ground ; this protects the graft during cold winter, and besides enables the plant to throw out roots of its own. They flower usually early in May.

"Waves Kissing Mountain." (*Kamadafuji.*) Very double large flowers, deep rose color at centre shading to light pink ; margin banded white.

"Purple Cloud." (*Kumoi-dsuru.*) Large double flowers of deep claret purple ; a novel and beautiful color.

"Torch Light." (*Adzuma-zaki.*) Magnificent double flowers of intense and brilliant vermilion scarlet.

"Flower Festival." (*Adzuma-nishiki.*) Large double flowers, deep carmine shaded with rose, edges of petals variegated with white and yellow.

"Bird of Red Plumage." (*Kumano-nishiki.*) Large double flowers, of a deep ox-blood red ; foliage green mottled with yellow and rose.

"Waves of Moonlight." (*Ayanishiki.*) Large double flowers, creamy white flaked with purest satiny white bars ; foliage mottled yellow and green.

Price for any of the above 6 varieties $1.50 each, or the set of 6 for $8.00. (*Can only be sent by express at purchaser's expense.*)

Mammoth Semi-Double Japanese Tree Pæonias.

Immense semi-double flowers measuring from 12 to 14 inches across. A large bunch of yellow stamens in the centre of each flower—like a golden coronet—greatly adds to the charming effect.

"Dancing in the Sun." (*Abaskinata.*) Monstrous flowers of rosy pink shading to blush white at margin ; interior base of petals bright madder surrounding large yellow centre.

"Brocade of Yezzo." (*Hana-tachibana.*) Immense flowers of blush white, lower half of petals striped rose ; yellow centre.

"Flying Dragon." (*Adzuma-Kagami.*) Magnificent large flowers of deep rich carmine-red—yellow centre.

Price for any of the above 3 varieties, $1.50 each, or the set of 3 for $4.00. (*Can only be sent by express at purchaser's expense.*)

RANUNCULUS.

Among dwarf flowers these are unrivaled for lovely form and bright and attractive colors, ranging through gorgeous shades of white, crimson, yellow, purple black, many of them being beautifully marked with other shades. They flower profusely in pots in the house or if grown in frames in the spring. (*See cut.*)

Double Turban Mixed. Peony-formed flowers, large and early, vivid colors. 2 for 5c., 20c. per doz., $1.00 per 100.

Double Giant French Mixed. Remarkable vigorous growers with immense and gorgeous flowers. 2 for 5c., 20c. per doz., $1.00 per 100.

Double Persian Mixed. Camellia or rose-shaped flowers, very double, rich variety of colors. 2 for 5c., 20c. per doz., $1.00 per 100.

RANUNCULUS.

6 Bulbs of one variety sold at dozen rates ; 25 at 100 rates. Bulbs are delivered free in the U. S., except were noted.

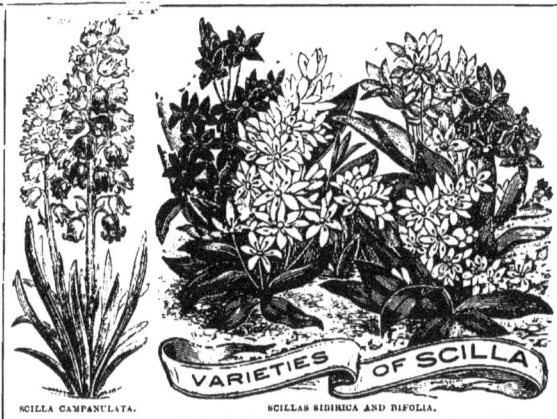

SCILLA.

Sibirica. (*Amana or Prccox.*) One of our most beautiful hardy spring bulbs, producing in profusion masses of exquisite rich blue flowers almost before the snow has disappeared. If grown in masses, their flowers fairly carpet the ground; and if grown with Snowdrops and Crocus, for contrast, the effect is magnificent. They should be largely planted as undergrowth in Hyacinth beds, etc. Grown in pots, they may be had in bloom from Christmas until April. (*See cut.*) 3 for 5c., 15c. per doz., $1.00 per 100, $8.00 per 1,000.

Bifolia. Most beautiful bright blue flowers borne on short spikes; these should be extensively planted; they are also fine for forcing. (*See cut.*) 2 for 5c., 20c. per doz., $1.25 per 100, $9.00 per 1,000.

Taurica. A new and valuable variety just discovered in the Taurus Mountains, in Asia Minor. It is the earliest of the early Scillas; the flowers, of the richest deep gentian blue, are of superior form, and borne from 10 to 20 on spikes often 8 inches long. The plant and foliage is larger than the type, and it grows well and flowers early. 5c. each, 50c. per doz., $3.50 per 100.

SCILLA CAMPANULATA.   SCILLAS SIBIRICA AND BIFOLIA.

SCILLA—(Continued).

Campanulata. (*Wood Hyacinth.*) Flowers borne on tall spikes, 1 to 2 feet high. Each flower measures nearly one inch across and droops gracefully; perfectly hardy; also fine for pot culture. (*See cut.*) We offer the following colors: Campanulata Blue, White, Rose. 3 for 10c., 25c. per doz., $1.75 per 100.

Hyacinthoides. (Nutans.) The Spanish Hyacinth. Fine blue, very showy, free bloomer. 3 for 12c., 30c. per doz., $2.00 per 100.

Peruviani or Clusi. (The Peruvian Hyacinth or Cuban Lily.) Very beautiful, blooming rather late in spring, bearing large pyramidal spikes of flowers, which remain in flower a long time; beautiful objects when grown in pots. Rich ultramarine blue. Not hardy. 15c. each, $1.50 per doz.

SNOWDROPS.

In the early spring months there is nothing more beautiful than a sheet of the snowy, graceful blossoms of the Snowdrop. Beds and effects of surpassing beauty may be arranged with Snowdrops in the centre, edged with bright blue Scilla Sibirica, or Chionodoxa Lucilie, or by intermingling them. When practicable, each planting of the Snowdrops should be permanent. The Snowdrop and the Crocus, when planted in alternate circles, are very effective, and follow each other so closely that no gap is left in the succession. In beds of Tulips and Hyacinths, Snowdrops are very effective between the lines; they flower while these bulbs are just moving the surface, and when the flowers are over there remains an elegant groundwork of green foliage. Used as a permanent edging, and in masses on the edges of lawns, nestling in the grass, they look charming.

Single Snowdrops. 3 for 5c. 15c. doz., 85c. 100, $6.00 1,000.

Double Snowdrops. 3 for 10c., 30c. doz., $2.25 100.

Elwes' Giant Snowdrops. One of the finest of the genus, at least three times the size of the ordinary single Snowdrop; flowers slightly marked with green spots; very fine for cutting and slightly sweet-scented. (*See cut.*) 3 for 5c., 15c. doz., $1.00 100, $8.00 1,000.

King of Snowdrops. (*Cassaba Robusta.*) See Novelties, page 8.

GIANT SNOWDROPS   SPARAXIS.

SPARAXIS.

Exceedingly large and beautiful blooms of about two inches across, of the most telling combinations, and of the brightest shades of color, certain to please even the most fastidious; they are tigered, blotched, spotted, streaked and flushed in the most diverse and pleasing manner. They are not hardy, but do exceptionally well when grown in the conservatory or house in pots or in cold frames. (*See cut.*)

Mixed Varieties. 2 for 5c., 20c. per doz., $1.25 per 100.

COPYRIGHT 1892 BY PETER HENDERSON & CO.

Bulbs delivered free in the U. S., EXCEPT WHERE NOTED. AVAIL YOURSELF OF PREMIUM. SEE SECOND PAGE OF COVER.

SPRING SNOWFLAKE.

Spiræa, or Astilbe. *Ready for shipment in November.*

Japonica. In garden culture it flowers freely during the summer, and is perfectly hardy, but its great value is when grown in pots for window and greenhouse decoration, and it is indispensable for forcing for cutting. The flowers are borne in large, feathery panicles of white, and last a long time in bloom. One of the finest plants for winter and spring blooming. Fine clumps, 15c. each, $1.50 per doz.; or, buyer paying transit, at $6.00 per 100.

Aurea Reticulata. Flowers pure white, in large clusters, foliage beautiful green, elegantly veined with yellow, very handsome. 25c. each, $2.50 per doz.

Nana Compacta Multiflora. Its merits consist in compact growth, ample foliage of brilliant green, and its wonderfully free production of feathery white flowers, borne in plume-like panicles of magnificent proportions. 20c. each, $2.00 per doz.

Astilboides floribunda. (See *Novelties*, page 10.)

STERNBERGIA LUTEA.

Schizostylis. *(Kaffir Lily, or Crimson Flag.)*

Coccinea. A very pretty, half-hardy bulbous plant; the leaves are neat and glossy and the flowers are rosy scarlet, borne on tall spikes; the bulbs are usually planted out in spring, and in the autumn the plants are lifted, potted and brought into the conservatory, where they will bloom for months; the more flowers that are cut from it the more spikes are produced. 3 for 12c., 35c. per doz., $2.60 per 100.

Spring Colchicum. *(Bulbocodium Vernum, or Meadow Saffron.)*

A charming early, spring-blooming plant, in flower two weeks before the Crocus, producing masses of rose-purple flowers, very beautiful for edgings and patches here and there. Breaking up through the snow, in juxtaposition with Snowdrops, it is a charming sight. Clumps of them dug up and potted in the winter and placed in a sunny window will soon be a mass of bloom. 3 for 12c., 35c. per doz., $2.50 per 100.

TRILLIUM GRANDIFLORUM. *(See description on opposite page.)*

Spring Snowflakes. *(Leucojum Vernum.)*

These produce flowers like monster Snowdrops, having the delicate fragrance of the Violet. It is one of our earliest spring flowers, with white blossoms distinctly tipped green; handsome in outline and prized for bouquets. Very graceful in growth, they should be grown in quantities in the borders or in clumps on the edges of shrubbery, where they are beautifully effective, and when established produce enormous quantities of flowers. They can also be slowly forced in pots for winter bloom. (*See cut.*) 3 for 10c., 30c. per doz., $2.00 per 100.

Sternbergia. *Mt. Etna Lily, or Lily of the Field.*

One of the most charming and useful of autumnal flowering bulbs. The flowers, which are produced from September to November, are much like a Crocus, but larger, and the petals more fleshy and of such firm texture that they withstand any amount of bad weather, brightening up our gardens long after other flowers are gone. They are not only very hardy, but increase rapidly. (*See cut.*)

S. Lutea. Large, pure yellow Crocus-like flowers, which are produced with the leaves during late autumn. This is supposed by some writers to be the true Lily of Scripture, as it grows abundantly in the vales around Palestine, etc. 3 for 10c., 30c. per doz., $2.00 per 100.

S. Macrantha. (See *Novelties*, page 9.)

Six bulbs of one variety sold at dozen rates, 25 at 100 rates. Bulbs delivered free in the United States, except where noted.

TROPÆOLUM JARRATTII.

- - TRILLIUM. - -

Grandiflorum. (Great American Wood Lily. This is one of the most beautiful American plants, perfectly hardy, growing and flowering profusely in partially shady nooks about the lawn, under trees, etc. The flowers are large, of the finest white, changing in a few days to soft rose; if grown several in a pot it makes one of the best white winter flowers. (See cut on page 46.) 5c. each, 50c. per doz., $3.75 per 100.

Sessile Californicum. A beautiful variety with mottled foliage. Flowers pure white. 10c. each, $1.00 per doz.

Johnsonii. New pink variety. See Novelties, page 9.

- - TRITONIA. - -

Exceedingly bright and free-blooming bulbous plants, highly valuable for both garden and pot culture. A dozen roots in a ten-inch pot will in the autumn make a beautiful display. The bulbs should be grown in pots in a cold frame during winter, and they can either be brought in the conservatory toward spring for blooming or the bulbs can be kept dormant and planted out in May, like Gladiolus, and then be lifted in the autumn for winter blooming.

Mixed Colors. 3 for 10c., 30c. per doz., $2.00 per 100.

WINTER ACONITE.

THE GOLDEN "WINTER ACONITE."

(Eranthis Hyemalis.)

Early in spring the golden blossoms of Winter Aconites look charming, resting on an emerald green cushion of leaves and forming a striking contrast to the pure white Snowdrop, Spring Snowflake, and the lovely blue Scillas and Chionodoxa. The Winter Aconite's flower appears in the garden very early. The large golden blossoms often unfold among ice and snow. For winter blooming in pots it is a bulb which all should have; the large golden blossoms show to great advantage. It is perfectly hardy and sure to succeed either in the garden or in pots. 6 to 8 inches high. (See cut.) 3 for 10c., 25c. per doz., $1.50 per 100.

- TROPÆOLUM. -

Beautiful and graceful climbing plants for the conservatory or window garden, producing in the early spring months richly colored flowers. They can be grown in pots and trained over low trellises—or as a bracket plant in the window. (See cut.)

Jarrattii. Scarlet, yellow and black. 15c. each, $1.50 per doz.

- TRITELEIA. -

Uniflora. (Spring Star Flower.) A perfect little gem for either pot culture or for borders. Each bulb produces several pretty star-shaped flowers, one or two inches across, of a delicate milky white suffused with blue and emits the perfume of primroses. (See cut.) 2 for 5c., 20c. per doz., $1.00 per 100.

TRITELEIA.

- - - VALLOTA. - - -

Purpurea. (Scarborough Lily.) A valuable free summer and autumn bloomer; color, rich red. It does well planted in the open ground in May, and when in bud can be potted and removed for conservatory or window decoration, or it can be grown continuously in pots or tubs as it improves with age. Requires repotting but seldom and can be left to grow, blossom and increase for several years, as well-established plants when in flower are simply magnificent. We know of no plant that will give more continued delight than this. It is one of the few really good window plants. (See cut.) 30c. each, $3.00 per doz.

ZEPHYRANTHES.

"Zephyr Flowers" and "Flowers of the West Wind," among our most beautiful dwarf bulbous plants, very effective for planting out in masses in May, flowering with great profusion during the summer. They are also most suitable for pot culture; 6 to 12 bulbs clustered in a 6-inch pot in the autumn will give a fine display of bloom during the winter in the window garden or conservatory. 1 foot high. (See cut.)

Atamasco. White suffused with flesh color, hardy. We know of borders of it as far north as Connecticut that have stood the colds of several winters. 5c. each, 50c. per doz., $2.50 per 100.

Candida. Large-flowering, pure white. 3 for 10c., 30c. per doz., $2.00 per 100.

Rosea. Large flowers, 3 to 4 inches across, of the most exquisite rose-pink. 6c. each, 50c. per doz., $3.00 per 100.

Sulphurea. Large light yellow flowers, dark foliage. 5c. each, 50c. per doz., $3.00 per 100.

VALLOTA PURPUREA.

ATAMASCO LILIES.

6 Bulbs of one variety sold at dozen rates; 25 at 100 rates. DELIVERED FREE IN THE UNITED STATES, except where noted.

Do not forget OUR LIBERAL PREMIUMS offered on second page of cover.

❀ ❀ ❀ LILIES. ❀ ❀ ❀

Lilies are matchless amongst hardy plants for beauty of form and variety of color. They commence flowering in May, and with the various species maintain a continuous and unbroken succession of bloom until autumn. No class of plants capable of being cultivated out-of-doors possesses so many beautiful; their stately habit, rich and varied colored flowers, often deliciously fragrant, and easy culture, render them so distinct from all other hardy plants that no collection, however select, should be without a few of the choicest sorts. They are also excellent subjects for the shrubbery border, if planted in groups between hardy Azaleas, Rhododendrons, etc., the soil suitable for these plants being particularly adapted for the growth of many kinds of Lilies, more especially our native and California species. The best time to transplant the bulk of the Lilies is in October and November.

All Lily Bulbs are delivered free in the United States at the single and dozen price, but at the 100 rate purchaser pays transit.

Orders for Lilies cannot be filled before **November,** *excepting for Candidum, Harrisii and Longiflorum, which are ready August 1.*

COLLECTIONS OF LILIES.

Collection of "Popular" Lilies.
Contains 1 bulb each of the following 12 varieties, viz.:

Auratum,	*Lancifolium album,*
Candidum,	*Longiflorum,*
Canadense,	*Martagon,*
Croceum,	*Philadelphicum,*
Elegans,	*Pomponicum,*
Lancifolium roseum,	*Tigrinum Double.*

Price, $1.75, *delivered free in the United States.*

Collection of "Choice" Lilies.
Contains 1 bulb each of the following 12 varieties, viz.:

Auratum Pictum,	*Humboldtii,*
Columbianum,	*Krameri,*
Chalcedonicum,	*Melpomene,*
Elegans Plena,	*Leichtlinii,*
Excelsium,	*Washingtonianum,*
Harrisii,	*Tenuifolium.*

Price, $5.00, *delivered free in the United States.*

Collection of "Rare" Lilies.
Contains 1 bulb each of the following 9 varieties, viz.:

Auratum Platyphylum,	*Colchicum,*
" *Vittatum-*	*Dalmaticum,*
" *rubrum,*	*Incomparabile,*
" *Wittei,*	*Giganteum,*
Browni,	*Hansoni.*

Price, $6.00, *delivered free in the United States.*

LILIUM AURATUM ❀❀
...THE...
GOLDEN-RAYED JAPAN LILY.

LILIUM AURATUM, "THE GOLDEN-RAYED LILY OF JAPAN," AND ITS VARIETIES.

These are the grandest and most satisfactory Lilies in cultivation for garden decoration, strong thrifty growers, profuse bloomers, and will always succeed if any Lily will. If you can only have one Lily let it be an Auratum.

Auratum. (The Golden-rayed Lily of Japan.) The flowers are pure white, thickly studded with crimson spots, while through the centre of each petal runs a clear golden band. Fully expanded, the flowers measure nearly a foot across, are produced abundantly from June to October, and possess a most delicious fragrance. 3 to 5 feet. (See cut.) Price, **First Size** bulbs, 15c. each, $1.75 per doz.; or, buyer paying transit, at $10.00 per 100. **Extra-large Bulbs,** 25c. each, $2.50 doz.; or, buyer paying transit, $16.00 per 100. **Mammoth Bulbs,** 50c. each, $5.00 per doz.

Auratum Vittatum Rubrum. Probably the grandest Lily in cultivation. Magnificent flowers 10 to 12 inches across, clear waxy white with a burnished crimson stripe half an inch wide through the centre of each petal, and the rest of the flower beautifully spotted crimson. Highly perfumed. $1.00 each, $10.00 per doz.

Auratum Pictum. A very choice variety, pure white, with a band through each petal, one-half of which is red and the other half yellow, entire flower beautifully spotted crimson. 60c. each, $6.00 per doz.

Auratum Witteii. A magnificent Lily, immense flowers, widely expanded, color purest white with a wide yellow stripe through the centre of each petal, often grows 6 feet high, very free-blooming, frequently bearing from 30 to 40 flowers on a stem. $1.25 each, $12.00 per doz.

Auratum Platyphyllum. This is without a question one of the most wonderful Lilies in cultivation. The leaves are very long and broad, and the stems attain a height varying from 7 to 10 feet. The flowers are similar in color to Auratum, heavily spotted, but are much larger, the petals more overlapping, and of greater substance. Immense bulbs. 75c. each, $7.50 per doz.

6 bulbs of one variety sold at dozen rates, 25 at 100 rates. Do not forget to avail yourself of our Premium offer on 2d page of cover.

LILIUM CANDIDUM.

Lilies ... (Continued).

Batemania. A Japanese Lily, growing from 3 to 4 feet high and producing bright apricot-tinted flowers; blooms in July. 15c. each, $1.50 per doz.

Brownii. One of the most beautiful Lilies; immense trumpet-shaped flowers 10 inches in length; interior, pure white with chocolate colored anthers; exterior, brownish purple, easily grown in any light sandy loam; it is also well adapted for pots. (*See cut.*) 75c. each, $7.50 per doz.

Candidum. (Annunciation, Madonna or St. Joseph Lily.) The well-known Garden Lily, snow-white fragrant blossoms; it is also one of the best forcing Lilies for florists; 3 to 4 feet; blooms in the open ground about June (*See cut.*) 13c. each, $1.25 per doz., or, buyer paying carriage, at $5.00 per 100.

Canadense. Our beautiful native "Canadian Lily." Bright yellow marked with copious spots of red; blooms in June and July; 2 to 3 feet high. 12c. each, $1.25 per doz.

Canadense Flavum. A variety of above; color, pure yellow. 15c. each, $1.50 per doz.

Canadense Rubrum. Bright crimson with darker spots. 15c. each, $1.50 per doz.

Chalcedonicum. (*Scarlet Martagon.*) Bright scarlet recurved flowers; blooms in June; 3 feet. (*See cut.*) 85c. each, $8.50 per doz.

Colchicum. (*Monodelphum or Szovitsianum.*) Rich citron color, spotted with black; one of the best Lilies; 2 feet; blooms in July. 85c. each, $8.50 per doz.

Columbianum. One of the most elegant and graceful of the Lilies; it grows 3 to 4 feet high, bearing from 10 to 30 brilliant orange-red medium-sized flowers, beautifully spotted with crimson and very fragrant. 30c. each, $3.00 per doz.

LILIUM CHALCE-DONICUM

LILIUM KRAMERI.
(*See description on page 51.*)

Croceum. (*Bulbiferum.*) Broad funnel-shaped flowers of beautiful golden, slightly tinted scarlet; 3 to 6 feet; blooms in July. 15c. each, $1.50 per doz.

Dalmaticum. (True.) A very rare and beautiful species, growing about 5 feet high, bearing from 30 to 40 flowers of a deep rich crimson purple, almost black; easily grown and one of the hardiest. 75c. each, $7.50 per doz.

Elegans Erectum. (*Thunbergianum or Umbellatum.*) Very hardy and succeeds anywhere; flowers erect and borne in clusters; orange, spotted scarlet; 1½ feet; blooms in June and July. (*See cut.*) 15c. each, $1.75 per doz.

Elegans Bicolor. Red, striped and flaked yellow, and shot with lilac. 20c. each, $2.00 per doz.

LILIUM BROWNII

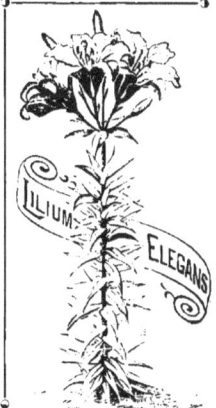

LILIUM ELEGANS

Elegans Aureum Maculatum. Light apricot yellow, spotted with black. 15c. each, $1.75 per doz.

Elegans Plena. Massive double flowers, deep reddish orange, spotted black; exterior, soft salmon color; thrifty and robust plant; 1½ feet high. 85c. each, $8.50 per doz.

Elegans Incomparable. The richest red known in Lilies; a deep ox-blood crimson, slightly spotted with black; very free and easily grown. 20c. each, $4.00 per doz.

Excelsum. (*Isabellinum or Testaceum.*) The stately form, beauty of color and delightful fragrance of this variety has made it a great favorite wherever known. It grows from 4 to 6 feet high, and produces from 6 to 12 nodding lilies of delicate light buff color; blooms in June and July. 50c. each, $5.00 per doz.

Giganteum. A splendid species of gigantic growth and very distinct. The stems grow 6 to 10 feet high, and frequently bear 12 to 20 large flowers 5 to 6 inches long. Color, creamy white, with a purple throat. The bulbs are very large; blooms in the summer. $1.25 each, $12.00 per doz.

The TRUE BERMUDA EASTER LILY. LILIUM HARRISII.

THE True Bermuda Easter Lily,

The Grandest Winter-flowering Plant in Cultivation. · · ·

This Peerless Lily is the greatest acquisition to floriculture made in many years. The remarkably short time required to bring them into flower, and the certainty to produce a profusion of bloom, and also the ease with which they can be made to flower at any desired period, such as Christmas, Easter and other special occasions, render them invaluable.

The Flowers are delightfully fragrant, pure waxy white, of great substance, and if cut as soon as they are open, or partially open, they can be kept for two weeks.

A Short Time only is Required to bring them into Bloom. Bulbs potted in August can be had in flower in November, if desired.

A Succession of Bloom can be kept up from November to May by bringing the pots of bulbs in from cold frames at intervals throughout the winter.

The Quantity of Bloom Produced is Marvelous; the average production of bulbs 5 to 7 inches in circumference, even when forced, is from 5 to 8 flowers, and of bulbs 7 to 9 inches, 8 to 12 flowers; and, if desired, a second crop of bloom, frequently equal to the first, can be secured the same season by drying the plants off after blooming for a time and then again starting them.

THE True Bermuda Easter Lily

Is of the Easiest Culture. Blooming in Wonderful Abundance. · · ·

Blooming Plants in Pots form typical Easter offerings and presents for decoration of the window, table, house or church. Nothing is more appropriately beautiful and highly appreciated. Thousands upon thousands are sold in the large cities every winter for this purpose.

As a Garden Lily it is of great beauty, being entirely hardy, excepting in northern climates, where it requires a protection of leaves or litter to the depth of 3 or 6 inches.

The TRUE Bermuda Easter Lily is our Great Specialty. We were the pioneer introducers of it and have always been "Headquarters," supplying the trade generally and the large dealers both in Europe and America.

Our Bulbs are Larger, Healthier and Better than those usually sold—for the reason that no expense is spared in manuring and cultivating. Experience having shown us that bulbs so treated are incomparably superior to bulbs grown on impoverished and underfertilized land year after year, which constitutionally weakens the bulbs and they give results accordingly. This is the general practice —especially among those who endeavor to grow cheaply—to enable them to undersell.

Full Cultural Instructions, telling how to force them into flower for Christmas or Easter—the treatment after flowering—about planting in the open garden, etc. Sent Free to every purchaser from this catalogue.

PRICES OF THE TRUE BERMUDA EASTER LILY BULBS.

(Now Ready for Shipment.)

SIZE OF BULBS.	Each. Delivered free in the U. S.	Per doz.	Per hundred. Express, buyer paying carriage	THE BEST SIZES TO USE.
Second Size Bulbs. (5 to 7 inches in circumference.)	10c.	$0.85	$5.00	The best for "Earliest" forcing, and particularly for cutting with long stems. Bulbs should produce from 5 to 8 flowers.
First Size Bulbs. (7 to 9 inches in circumference.)	15c.	1.50	8.00	This is the most popular size to grow as pot plants for decoration and equally as valuable for cutting. Bulbs should produce 8 to 12 flowers.
Extra Size Bulbs. (9 to 11 inches in circumference.)	25c.	2.50	15.00	These extra bulbs are usually grown for specimens in pots. Each bulb should produce from 12 to 18 flowers.
Monstrous Bulbs. (11 to 14 inches in circumference.)	60c.	6.00	40.00	These monstrous bulbs make fine specimens for exhibition and decoration. As many as a hundred flowers have been produced from one bulb.

6 furnished at dozen rates, 25 at 100 rates. 1,000 rates on application. Delivered in U. S. free, except where noted.

LILIES—(Continued.)

LILIUM HANSONI.

Hansoni. (*The True Japanese Yellow Martagon.*) This is one of the best of the Martagon group, unsurpassed in vigor of growth and beauty by any other Lily. The flowers are large, of a bright yellow, tinged orange and spotted maroon, with thick wax-like petals. Very robust, free and easily grown. (*See cut.*) 85c. each, $8.50 doz.

Humboldtii. (*Bloomerianum.*) A remarkably fine variety, attaining the height of from 3 to 5 feet and producing freely large flowers of golden yellow color, spotted with purple. A native of California. July. 35c. each, $3.50 doz.

Krameri. Distinct from all other Lilies ; large flowers of a soft, beautiful rose color. Large bulbs, 50c. each, $4.00 doz. (*See cut, page 49.*)

Longiflorum. Well-known beautiful, snow-white, fragrant, hardy garden Lily ; flowers trumpet-shaped, 6 to 8 inches long. This is also a splendid variety for forcing for winter flowers, and is also known as the "Easter or St. Joseph Lily." Height, 1½ to 5 feet. In the open ground it blossoms in June and July. (*See cut*)

First size, 5 to 7 bulbs, each, 10c. ;	dozen,	$1.00
Extra " 7 to 9 "	"	15c. ;	"	1.50
Mammoth,9 to 11 "	"	25c. ;	"	2.50

LILIUM LONGIFLORUM.

Lancifolium or Speciosum. These varieties are perhaps the most popular Lilies grown, being easy of cultivation and produce large flowers of delicate beauty on strong stems. Height from 2 to 5 feet; blooms in August. (*See cut.*) The varieties are:

L. Album. Purest white. 1st size, 20c. each, $2.00 doz.; extra size, 30c. each, $3.00 doz.

L. Roseum. White, shaded and spotted with rose. 1st size, 15c. each, $1.50 doz ; extra size, 25c. each, $2.50 doz.

L. Rubrum. White, shaded with deep rose and spotted red. 1st size, 15c. each, $1.50 doz. ; extra size, 25c. each, $2.50 doz.

L. Melpomene. Flowers very large and very abundant, of rich blood-crimson, heavily spotted. 1st size, 20c. each, $2.00 doz.; extra size, 30c. each, $3.00 doz

Leichtlinii. A beautiful Japanese species, of neat and elegant habit ; the flowers are pure canary yellow, with crimson spots A real acquisition. 3 to 5 feet; blooms in August. 50c. each, $5.00 doz.

Martagon. Purplish red, spotted with dark purple. Prolific bloomer. 2 to 3 feet high. Blooms in midsummer. 25c. each, $2.50 doz.

LILIUM LANCIFOLIUM

Pardalinum. Scarlet, shading to rich yellow, freely spotted with purple brown. 3 feet. July and August. (*See cut*) 20c. each, $2.00 doz.

Philadelphicum. Native variety; bright orange-red, spotted with purple. Height, 1 to 3 feet; blooms in midsummer. 12c. each, $1.25 doz.

Pomponicum Rubrum. (Scarlet Turban Lily.) This is very early-flowering, growing about 3 feet, bearing numbers of fiery-scarlet flowers. It grows freely, is one of the most effective. 30c. each, $3.00 doz.

Superbum. (Turk's Cap Lily.) One of our native species. When established in good rich soil it will produce upwards of 50 beautiful orange, tipped red, spotted flowers in a pyramidal cluster. 3 to 5 feet (*See cut.*) Blooms in July. 15c. each, $1.50 doz.

Tenuifolium. A miniature Lily, having slender stems, 18 inches high, bearing 12 to 20 fiery-scarlet flowers ; a gem for cutting and easily grown in pots or in a warm dry border, one of the first in bloom. 30c. each, $3.00 doz.

Tigrinum. The old Favorite Tiger Lily Large flowers, orange red, spotted with black ; hardy, free, and always does well. 12c. each, $1.25 doz.

Tigrinum Splendens. (Improved Tiger Lily.) The grandest of the Tigers, black polished stem, sometimes 6 feet high. Very large pyramids of flowers, orange-red, spotted with black. 3 to 5 feet; blooms in August. 15c. each, $1.50 doz.

Tigrinum Flora Plena. (Double Tiger Lily.) This is a plant of stately habit, growing from 4 to 6 feet high, bearing an immense number of double bright orange red flowers, spotted with black ; blooms in August. 15c. each, $1.50 doz.

Wallacei. Flower, rich vermilion orange, spotted with raised maroon dots, autumn flowering. thrifty grower, and highly satisfactory. 15c. each, $1.50 doz

Washingtonianum. A beautiful variety from Oregon, growing stiff and erect; flowers white, tinted with purple and lilac; 8 to 9 inches across when fully expanded. 3 to 5 feet high; blooms in the summer. 35c. each, $3.50 doz.

LILIUM PARDALINUM.

LILIUM SUPERBUM.

LATANIA BORBONICA.
For description and price, see page 54.

PLANT DEPARTMENT

Pages 52 to 63, inclusive.

PLANTS FOR CONSERVATORIES. .
PLANTS FOR GREENHOUSES. . . .
PLANTS FOR WINDOW GARDENS.

| HARDY SHRUBS, | | STRAWBERRIES |
| HARDY ROSES, | | AND GARDEN FRUITS. |

PALMS. We call particular attention to our list of **Palms**, and venture to say that for the money expended there is nothing we offer which will give more satisfaction to our customers. They are especially grown for window and room culture, and with ordinary care can be kept in perfect condition for years. Our stock of Roses, Carnations, Primroses, Chrysanthemums and other winter blooming plants is exceptionally fine, and contains only such varieties as are specially adapted for this purpose. In this connection we would say that **the variety of plants** which will thrive and bloom during the winter months is comparatively limited, and the lack of success which occasionally discourages amateurs is due more frequently to a poor selection of varieties than to mistakes in culture. With this point in view, we have carefully selected only such sorts as are adapted for this purpose, so that the most inexperienced amateur is perfectly safe in ordering from this catalogue, from the fact that it contains only such plants as are highly ornamental, either in flower or foliage, during the fall, winter and early spring months. We also offer a range of sizes in nearly everything, thus enabling customers to secure a large plant for immediate effect, or a small one at a low price, if desired.

SAFE ARRIVAL ASSURED.

We guarantee the safe arrival of all plants *sent by express*. Should anything be injured in transit notify us at once on receipt of the goods, and send us a list of what has been damaged, so that we can replace them without delay. But when they arrive in good condition our responsibility ceases, and if from inattention or other causes they fail and complaint is then made, we cannot replace them. *Plants* sent other than by express are entirely at the risk of the purchaser.

NOTICE.—Unless otherwise ordered we send all plants by express. Hardy shrubs and vines and hardy roses may be sent safely by freight, but unless the shipment is a large one express is cheaper, quicker and safer. Small lots of Geraniums, Fuchsias, Young Palms, Begonias, Ferns, Violets, Heliotropes and Primulas may be sent free by mail, if desired.

Have your Plants sent by Express.
Under the new ruling of the leading Express Companies, advocated and secured by us, plants packed in closed boxes or baskets will now be carried at a reduction of 20 per cent. from the regular merchandise rates.

COLLECTIONS OF PLANTS FOR HOUSE CULTURE.

Many amateurs are often at a loss what plants to select from the great variety offered in a catalogue, and to aid such we offer the following collections. In these collections we send our best plants and finest varieties, and purchasers secure them at less than regular rates, but in every instance the *selection* must be left to us. These collections embrace all the leading plants suitable for winter-flowering or decorating, all of which may be grown in a light window or sitting-room, such as Abutilons, Azaleas, Begonias, Carnations, Chrysanthemums, Fuchsias, Geraniums, Heliotropes, Palms, Primulas, Roses, Violets, etc., etc.

Collection No. 1.	12 plants for	$2.25
Collection No. 2.	25 plants for	4 00
Collection No. 3.	50 plants for	7 00
Collection No. 4.	100 plants for	12 00

See " Window " Collection of Palms, next page.

T꜓ "WINDOW" COLLECTION OF PALMS.

THESE we offer at very low prices for the collection so as to induce our customers to procure them, believing that there is nothing we offer which will give greater satisfaction for the money invested. These young Palms are easily grown into large specimens—one year's growth increasing their value fourfold at least. As some will, no doubt, want larger plants we offer them in **two sizes**; the larger size will make a fine display at once as they all show **character leaves.** We can send the second size collection *free by mail*, if desired, but we advise that they be **sent by express** (buyer to pay charges), as in that case we can send larger plants and will leave all the soil on the roots, an undoubted advantage. The first size we cannot send by mail. Below we give short descriptions of the plants.

LATANIA BORBONICA. (Fan Palm.)

The typical Palm and recognized as being indispensable in every collection. The second size of this plant will not have the fan-shaped—also called "character"—leaves; they develop the second year. First size, 75c. each; second size, 25c. each.

ARECA LUTESCENS. (The Ostrich Feather Palm.)

One of the grandest and most useful Palms in cultivation, full of grace and beauty. First size, 75c. each; second size, 30c. each.

KENTIA BELMOREANA. (The Curly Palm.)

The finest perhaps of all Palms for house culture, and very beautiful in form. First size, 75c. each; second size, 35c. each.

SEAFORTHIA ELEGANS. (The Giant Palm.)

Invaluable on account of its rapid growth and gracefully arched foliage. In its native state this is one of the most imposing Palms known and makes a fine specimen early in cultivation. First size, 50c. each; second size, 30c. each.

DRACENA INDIVISA. (The Fountain Palm.)

This plant is unsurpassed for hardiness in the house and eminently fitted to contrast with Palms and other decorative plants. First size, 50c. each; second size, 25c. each.

GREVILLEA ROBUSTA. (The Silk Oak.)

A magnificent plant for decorative purposes, of rapid, easy growth, finely cut foliage, rivaling a rare Fern. The young growths are a light bronze color, the tips being covered with a soft down closely resembling raw silk, hence the name of "Silk Oak." First size, 50c. each; second size, 25c. each.

CYPERUS ALTERNIFOLIUS. ... (The Umbrella Plant.) ...

Although, strictly speaking, not a Palm, yet its habit and appearance entitle it to a place among them. It is happily styled "The Umbrella Plant," as the leaves radiate from the stem and curve downward in graceful fashion. Of the easiest culture, a beautiful living green, and will prove satisfactory in any collection. First size, 50c. each; second size, 25c. each.

►►►►►►►►►►► PRICES OF THE "WINDOW" COLLECTION OF PALMS. ◄◄◄◄◄◄◄◄◄◄◄

First size,   7 strong plants from 4 and 5 inch pots.....................$3 00 (if bought separately would cost $4.25).
Second size, 7   "   "   " 2 and 3   "   .... 1.00 ("   "   "   "   1.95).

Pamphlet of Instructions on the care of Palms in dwelling-houses sent free to purchasers of either of the above collections.

WINDOW Decorated with PALMS

Palms and Decorative Plants.

KENTIA BELMOREANA.

LATANIA BORBONICA (Fan Palm).

This fine Palm is too well known to need any extended description. (*See cut, page* 52.) Price, 3-in. pots, 25c. each, $2.25 per doz.; 4-in. pots, 40c. each, $4.00 per doz.; 5-in. pots, $1.00 each, $9.00 per doz.; 6-in. pots, $1.50 and $2.00 each; 7-in. pots, $3.00 each.

KENTIAS BELMOREANA AND FOSTERIANA.

The Kentias are among the best of the Palm species for general cultivation. (*See cut.*) Price, strong plants, 3 feet high, $3.50 each; second grade, fine plants, $2.00 each; plants from 5-in. pots, $1.00 each, $9.00 per doz.; from 4-in. pots, 50c. each, $4.50 per doz.; from 3-in. pots, 35c. each, $3.50 doz.

LIVISTONA ROTUNDIFOLIA.

One of the prettiest Palms in cultivation, especially suited for table decoration. The foliage, which is similar to Latania Borbonica, but smaller, is gracefully recurved so as to form almost perfectly round plants; in fact, it might be called a miniature Latania. Fine plants from 5-in. pots, 4 to 5 *character* leaves, $2.00 each.

COCOS WEDDELLIANA.

This beautiful Palm is unquestionably the most elegant and graceful in cultivation. The finely cut leaves are recurved with exquisite grace. Price, 12 to 15 in. high, 75c. each; smaller plants, 35c. each, $3.50 per doz.

ARAUCARIA EXCELSA.

(*Norfolk Island Pine.*)

Deep green, feathery foliage, arranged in whorls, rising one above the other at regular distances; its symmetry of form, grace and beauty of foliage are unequaled in the vegetable kingdom. (*See cut.*) Price, 24 ins. high, $4.00 each; 15 to 18 ins., $2.50 each; 10 to 12 ins. high, $1.00 each.

PANDANUS UTILIS.

ARECA LUTESCENS.

This majestic Palm is without a peer for strength and elegance combined. Price, young plants from 3-in. pots, 30c., $3.00 per doz.; from 4-in. pots, 50c. each, $5.00 per doz.; from 5-in. pots, $1.00 each, $9.00 per doz. Specimen plants, $2.00, $3.00 and $4.00 each, according to size.

RUBBER PLANTS.

We have a splendid stock of this useful and ornamental plant, which will flourish under the most adverse conditions. Price, 12 to 15 ins. high, 75c. each, $7.50 per doz.

PANDANUS UTILIS.

Pandanus is perhaps the most useful of our ornamental foliage plants. As a vase plant or single specimen in greenhouse or conservatory it cannot be surpassed. (*See cut.*) Price, 7-in. pots, $2.00 each.

DRACÆNA INDIVISA.

From its gracefully drooping habit it sometimes is called the "Fountain Plant." Price, 50c. each; small plants, 25c. each.

DRACÆNA FRAGRANS.

Deep green, broad, gracefully drooping leaves. Price, plants from 5-in. pots, $1.00 each, $9.00 per doz.; 4-in. pots, 60c. each, $6.00 per doz.

DRACÆNA TERMINALIS.

Bronze red, variegated crimson and pink. 40c. each, $4.00 per doz.

DRACÆNA LINDENII.

One of the finest of the species. Broad strong leaves, gracefully recurved, heavily banded with creamy white and yellow on deep green ground. 6-in. pots, $2.00; 5-in. pots, $1.50; 4-in. pots, $1.00 each.

DRACÆNA STRICTA-GRANDIS.

A very handsomely colored variety, broad leaves, richly variegated crimson and pink, on a deep bronze. Price, 4-in. pots, 50c. each.

PANDANUS VEITCHII.

This is a grand decorative plant, creamy white variegation. Fine plants, 6-in. pots, $2.00 each; 8-in. pots, $4.00 each.

SEAFORTHIA ELEGANS.

A quick-growing, robust Palm full of grace and strength; the foliage is a rich green and is gracefully arched. It makes a fine specimen plant in a short time. Price, fine plants from 5-in. pots, 75c. each, $7.50 per doz.; smaller plants from 3-in. pots, 30c. each.

CYPERUS ALTERNIFOLIUS GRACILIS.

This is a new variety of the popular "Umbrella Plant," with very narrow foliage, which makes it much more desirable. It is most useful, either as an aquarium plant or as an ordinary house plant, succeeding under either culture. Price, fine plants from 5-in. pots, 50c. each.

ARAUCARIA EXCELSA.

The *Crimson* Rambler.

This grand Rose improves upon acquaintance.

The better it becomes known the more popular it becomes. . . .

Its Growth.

It is of rapid, vigorous growth; plants in our grounds attained a height of fifteen feet the past season. The flowers are produced in trusses, pyramidal in shape, good specimens measuring nine inches from base to tip, and seven inches across, fairly covering the plant from the ground to the top, so that it is **one mass of glowing crimson**. The color is superb and remains strong and vivid to the end; plants in our grounds retained the bloom for two months. The profusion of bloom is marvelous, **over three hundred blooms** having been counted on one shoot.

It is essentially a garden Rose, but makes a magnificent specimen in a pot or tub.

We offer fine pot-grown plants of this grand climbing Rose which, if planted this fall, ought to give a nice crop of bloom next season. These plants would also make good specimens in pots in the greenhouse or conservatory in the spring, if potted and started into growth early. Price, 60c. each, $6.00 per doz.; plants from open ground (ready October 25th), 40c. each, $4.00 per doz.

The *Memorial* Rose.

(Ready October 25th.)

.... (Rosa.... Wichuraiana.)

The grandest plant for cemetery decoration in existence. It grows flat on the ground like an ivy and blooms in grand abundance throughout July and intermittingly the balance of the season, pure white, sweetly fragrant flowers with a golden yellow disc. It is perfectly hardy, and if planted this fall should bloom next season. Price, extra strong plants, 40c. each, $4.00 per doz.

Winter....

Flowering Roses.

Bridesmaid. Brilliant pink.
Mme. de Watteville. Creamy, yellow edged rose.
Sunset. Rich orange and yellow.
C. de Noue. Brilliant carmine red.
Papa Gontier. Rich carmine.
Perle des Jardins. Deep yellow.
La France. Peach blossom pink.
The Bride. Pure white.
Mrs. P. Morgan. Bright rose pink.
Pres. Carnot. Rosy flesh pink, 40c. each, $4.00 per doz.
Meteor. Velvety crimson red.

Price, plants from 4-inch pots (except where noted), 25c. each, $2.25 per doz.; $16.00 per 100.

Note our Premium Offers on 2d page of cover.

This illustration shows a plant of the CRIMSON RAMBLER on the estate of Mr. Louis B. McCagg, Newport, R. I., taken July, 1896. The plant was set out in May, 1895, and was furnished by us. About 18 inches high when planted; when the photograph was taken it was **fifteen feet high and six feet wide**.

COPYRIGHT BY 1896 PETER HENDERSON & Co.

HARDY ROSES FOR FALL PLANTING FROM OPEN GROUND.

WE advise fall planting of Hardy Roses wherever the ground can be made ready, and we offer below a select list of sorts for that purpose. All the plants are from open ground, strong healthy plants, and will be ready for delivery about November 1st. *They are all budded low on Manetti stock.* Parties desiring to make large rose gardens this fall would do well to place their orders early if located north of New York City, so as to obtain early delivery.

Alfred Colomb. Carmine crimson, a grand Rose in every way.

Anna de Diesbach (Gloire de Paris). Rich carmine.

Baron de Bonstetten. Blackish crimson, with vivid red shadings.

Baroness Rothschild. An exquisite shade of satiny pink.

Coquette des Alpes. White, tinged blush, medium-sized flower, semi-cupped in form, a fine variety.

Eugene Furst. Velvety crimson, very large flower, with broad, massive petals quite double; a valuable Rose.

Gen. Jacqueminot. Brilliant crimson. The most widely popular Rose.

John Hopper. Bright rose with carmine centre, large and full; esteemed by all who grow it as of the highest order.

Louis Van Houtte. Crimson maroon, full and half-globular.

Margaret Dickson. White, with pale, flesh centre; extra large flower of fine substance, strong, vigorous growth.

Magna Charta. Dark pink, one of the easiest Roses to grow.

Mme. Gabriel Luizet. Light satiny pink, an attractive sort.

Mrs. John Laing. Rich, satiny pink, delicious fragrance; blooms constantly.

Mrs. J. Sharman Crawford. Deep rosy pink, very free-flowering.

Paul Neyron. Flowers 5 inches across; color, lovely dark pink.

PERSIAN YELLOW. Hardy *yellow* Rose; best of its color.

Prince Camille De Rohan. Dark crimson maroon, almost black.

Ulrich Brunner. Cherry-red; a grand Rose; very free-blooming. (*See cut.*)

Rosa Rugosa Alba. The white-flowered variety, ornamental foliage and showy fruit.

Rosa Rugosa Rubra. The red variety.

Price, for any of the above, 30c. each, $3.00 per doz.; full set of 20 sorts for $5.00.

ULRICH BRUNNER.

HARDY ....CLIMBING ROSES.

Baltimore Belle. Blush white.

Anna Maria. Large, rosy pink.

Empress of China. Bright pink.

Gem of the Prairies. Deep rosy carmine.

Price, 30c. each; set of 4 for $1.00.

OUR NEW GOLD MEDAL ROSE, JUBILEE.

Having purchased the entire stock of this grand Rose we control it absolutely. As a protection to our customers we will attach our lead Seal to every plant, and this seal must be attached to the plant to insure its genuineness.

THIS grand Rose was shown on a colored plate in our spring catalogue. It was awarded the gold medal by the Massachusetts Horticultural Society, the only Rose to obtain that honor. It is the finest dark red hardy garden Rose yet produced. A pure red in its deepest tone, shading to deep crimson-red and velvety maroon-red in the depths of the petals. It fairly glows in its rich warmth of coloring and has a rich velvety finish all its own, containing the darkest color combined with pure color of which nature is capable.

Price, strong plants from 5 inch pots, $2.00 each, extra strong plants from 6 inch pots, $3.00 each. These plants are two years old, grown on their own roots, and should produce a nice crop of flowers next year if planted this fall.

September Flowering Chrysanthemums

OUT of over 200 sorts which we tested we have selected those named below as the most desirable, and all bloomed the latter part of September and during early October. Hitherto the great objection to Chrysanthemums in our Northern States was that the frost destroyed most of the flowers before coming to maturity.

Baron Veillard. Brilliant yellow, lined rosy crimson.
Camille Bernardin. Very large peony-flowered; amaranth violet and brilliant carmine, lightened with white; entirely new. (*See cut.*)
Charles Joly. Beautiful violet rose.
Chas. Greard. Yellow ground, marked tawny red.
Chev. Ange Bandiera. White, cream and rose, golden centre.
Jean Nicolas. Japanese; petals slender and curled; glossy rose and a shade of dim white, centre cream. (*See cut.*)
Marquise de Montmort. Glossy rose and silvery white.
Mlle. Fleurot. White, tipped rose, yellow centre.
Mlle. Jacob. Japanese; recurved, rosy lilac and glossy white.
Mlle. Germaine Cassagneau. Lilac rose, shaded with white.
Mme. A. Thiebault de la Croure. Carmine purple, centre yellow.

Mme. Gastelier. Pure white, a large fluffy flower, double to the centre; one of the earliest to bloom. (*See cut.*)
Mme. Mathilde Bettzich. White, marked light rose.
Mme. Ve. Pasquier. Creamy white, shaded rose.

Mme. F. Bergmann. Pure white; a fine large flower.
Mrs. Chas. W. Woolsey. Pure white; very early.
M. Valery Larbaut. Creamy white and rose, lemon centre.
Vve. Chiquot. Brilliant yellow, red centre.

Price for any of the above, 30c. each. Set of 18 sorts for $4.50.

☞ *All the plants offered above are growing in 5-inch pots and are in bud and bloom.*

Our pamphlet on the culture of the Chrysanthemum will be sent free to buyers from this catalogue.

NOTE.—These plants offered above cannot be sent by mail; they can only be sent by express, buyer to pay charges.

....EXHIBITION COLLECTION....

This collection, as its name implies, is made up of such sorts as are generally grown for exhibition purposes, and embraces nearly every shade of color and variety of form known in the Chrysanthemum.

Daisy. A *single* flowering variety; clear white with yellow centre; very free-blooming and very pretty.
Ivory. Pure white, dwarf branching habit; it is considered the finest sort in its color.
John Shrimpton. Very dark crimson.
King of Ostrich Plumes. Deep chrome-yellow, shaded buff and orange, large and double, perfectly incurving; long broad petals, covered with glandular hairs.
L'Enfant des Deux Mondes. (Ostrich Plume.) Magnificent white "sport" from L. Boehmer.
Louis Boehmer. "The Pink 'Ostrich Plume' Chrysanthemum." One of the best of the type.

Major Bonnaffon. Grand incurved yellow of largest size.
Maud Dean. Perfectly double; color, deep pink.
Miss Minnie Wanamaker. Large clear white; very fine.
Marion Henderson. This grand variety is invaluable for *early cut flowers.* Bright yellow, fine form.
The Queen. The flower excels all in pure whiteness; is extra large, broad and deep.
W. B. Dinsmore. Golden yellow; very attractive.
Wm. Falconer. (Ostrich Plume.) Light pink.
Viviand Morel. Light rose, creamy white and pink.
W. H. Lincoln. Bright yellow; large, splendid flower.
Yellow Queen. Extra large; bright yellow; very early.

Price for any of the above (except where noted), 30c. each, $3 00 per doz.—fine plants from 5-inch pots.

NOTE.—These cannot be sent by mail; they can only be sent by express, buyer to pay charges.

SPECIAL NOTICE.—Want of space in this catalogue will not permit us to list our full collection of Chrysanthemums. We have in stock, however, all the varieties offered in our spring catalogue of "EVERYTHING FOR THE GARDEN," and can supply them at the prices quoted therein—plants from 2-inch pots.

If the selection is left to us, we can furnish a grand assortment at $1.00 per doz., $6.00 per 100.

NOTE OUR OFFER OF PREMIUMS ON 2D PAGE OF COVER.

AZALEA INDICA. (Greatly reduced.)

ANTHEMIS CORONARIA FL. PL.

Double Golden Marguerite. Nothing can be more showy, either as a pot plant or bedded out. It is a perfect mass of rich golden yellow flowers the whole year round, being equally useful in winter as in summer. Price, 15c. each, $1.50 per doz.

ASPIDISTRA LURIDA VARIEGATA.

A beautiful plant with large, lance-shaped leaves, finely variegated with clear, cream-colored stripes. An elegant window or conservatory plant of the easiest culture. $1.00 each.

... ASPIDISTRA LURIDA ...

A green-leaved variety of the above, of strong growth; will succeed in any position; an excellent hall or corridor plant. 75c. each.

AZALEA INDICA...

We offer a very fine lot of Azaleas, comprising the most distinct and best varieties in cultivation, embracing all shades of crimson, white, pink and rose color. They are shapely specimens, well "headed," double and single, splendid plants for winter and spring decoration.

10 to 12 inch heads.........$0.75 $7.50
12 to 14   " 1.00   9 00
14 to 16   " 1.50   15.00

Extra large plants, 18 to 20 inch heads, $3.50 each; 20 to 24 inch heads, $5.00 each.

CAMELLIA JAPONICA. The collection embraces double white, double pink, double red and double variegated. Price, $1.00 each.

... ASPARAGUS ...

Greenhouse climbing plants of rare beauty; they are specially suited for window gardening.

Sprengeri. A desirable species, useful as a pot plant or for baskets; the fronds are frequently four feet long, a rich shade of green, retaining their freshness for weeks after being cut. A fine house plant, as it withstands dry atmosphere. 30c. each, $3.00 per doz.

Plumosus Nanus. (*Climbing Lace Fern.*) Bright green leaves, gracefully arched, and as finely woven as silken mesh, retaining their freshness for weeks when cut. 30c. each, $3.00 per doz.

Tenuissimus. Very fine filmy foliage. A handsome climbing plant for the window. 15c. each, $1.50 per doz.

BEGONIAS (in Variety). The flowers are beautiful, drooping in graceful panicles of various colors. Our collection is composed of standard sorts. 15c. each, $1.50 per doz.

CALCEOLARIA. Splendid plants for winter and spring blooming, large pocket-shaped flowers, exhibiting wonderful colors and markings. 15c. each, $1.50 per doz.

CINERARIA HYBRIDA. Magnificent flowering plants for greenhouse and home. Flowers borne in clusters on upright stalks, large handsome foliage. 15c. each, $1.50 per doz.

Carnation
Wm. Scott.

WINTER=FLOWERING CARNATIONS.

Alaska. Pure white, the leading variety for winter blooming.
Lizzie McGowan. One of the best *white* varieties.
Emily Pierson. Deep red, fine large flowers.
Daybreak. So called because of its rare and beautiful color, compared to the first faint tinge of rosy light seen in the eastern sky.
Rose Queen. Deep brilliant rose.

Portia. Brilliant scarlet.
Wm. Scott. Flowers large and non-bursting, of a deep Grace Wilder pink, stems long, of good strength. (*See cut.*)

Price, plants from open ground, *not potted*, 25c. each, $2.50 per doz., $16.00 per 100; strong plants established in 5 and 6 inch pots for winter blooming, 40c. each, $4.00 per doz., $25.00 per 100.

ADIANTUM CUNEATUM ("MAIDENHAIR" FERN).

THE "BOSTON" FERN. (Nephrolepis Bostoniensis.)

In well-grown specimens the fronds attain a length of 6 or 7 feet, like plumes arching over in every direction, in a most graceful manner. This beautiful Fern is excellent for outside planting in shady borders. It is a very fine plant for hanging pots or baskets on the piazza in summer and conservatory or window in winter. Price, 1st size, 75c. each; 2d size, 50c. each; 3d size, 25c. each, $2.25 per doz.

HOUSE FERNS.

Our collection contains the choicest of the "Maidenhair" varieties (*see cut above*) and best basket and vase sorts. They are of easiest culture, and nothing adds more grace and beauty to a home than a few handsome Ferns. 1st size, 25c. each, $2.50 per doz.; 2d size, 15c. each, $1.50 per doz.

SELAGINELLA EMILIANA.

A beautiful variety, fine as a Fern and excellent for pot culture. 25c. each, $2.25 per doz.

"PANSY" GERANIUM.

(The Old-fashioned Lady Washington Geranium.)

We take pleasure in being able to offer plants of this very useful Pelargonium. There is nothing that gives better returns for a little care, when grown in the window garden, than this plant; it will be literally covered with flowers. It is low and spreading in habit and very easily grown, requiring little attention. The coloring is very unique, a light pink with dark blotches. It is this combination of coloring with blotches which obtained for it the popular title of "Pansy Geranium." (*See cut.*) Price, 35c. each, 3 for $1.00, 7 for $2.00, 12 for $3.00.

WINTER-FLOWERING FUCHSIAS.

Jos. Rosain. Corolla plum color, sepals dark red.
Cervantes. Purple corolla, scarlet sepals.
Molesworth. White corolla, crimson sepals.
Rosains Patrie. White corolla, rosy carmine sepals, double.
Storm King. Corolla pure white, very free.
Trophee. Sepals clear red, corolla dark violet blue.
Beacon. Corolla a deep carmine, sepals scarlet.
Lottie. Corolla carmine, sepals white.
Speciosa. Corolla orange scarlet, sepals white.

Price, 10c. each; set of 9 Fuchsias for 75c.

DOUBLE GERANIUMS. (Winter Flowering.)

Young vigorous plants, which will produce flowers abundantly throughout the winter and can be planted out-of-doors next spring, blooming all season. Plants are from 4-inch pots.

La Favorite. Pure white.	**Silver Queen.** Purest white.
Golden Dawn. Orange scarlet.	**V. P. Dubois.** Deep scarlet.
Gloire de France. Salmon.	**Double Gen. Grant.** Bright red.
Beaute Poitevine. Rich glowing salmon, a grand variety.	**Le Cid.** Dark crimson.
Mme. Thibaut. Deep rose.	**Naomi.** Deep pink.
	De Lacepede. Light pink.

Price, 15c. each, $1.50 per doz., $1.25 per set of 10.

SINGLE GERANIUMS. (Winter Flowering.)

Marguerite Lyer. White.
Daybreak. Shell pink.
Jules Ferry. Crimson.
Romeo. Light pink.
Beauty of Ramsgate. Crimson.
Blazing Star. Brilliant scarlet.
E. Bergmann. Intense scarlet.
Mme. La C. de Pot. Salmon.

Price, 15c. each, $1.50 per doz.; set of 8 for $1.00.

HELIOTROPES (in Variety).

Indispensable for bouquets and vases of flowers. Their rich tints of lavender, blue and purple, and exquisite vanilla perfume, are familiar to all. 10c. each, $1.00 per doz.

"THE PANSY GERANIUM."

SINGLE WHITE CHINESE PRIMROSE.

CHINESE PRIMROSES. .

For winter flowers there is no more desirable plant than the Chinese Primrose. They are easily grown and flower incessantly throughout the winter. Our plants comprise all the shades of *crimson, pink, white* and *variegated.* 1st size, 25c. each, $2.25 per doz.; 2d size, 15c. each, $1.50 per doz. (*See cut.*)

VIOLET, MARIE LOUISE.

(*Winter Flowering.*)

The well-known double purple variety. We offer strong healthy plants, so that any amateur should have no difficulty in growing them. 20c. each, $2.00 per doz., $12.00 per 100.

HARDY ENGLISH VIOLET

(*Spring Blooming.*)

This Violet is entirely hardy, perfectly double, a deep violet color and most deliciously fragrant. It surpasses the well-known "Marie Louise" Violet in richness of color, being many shades darker, and far excels it in its delightful odor; this is one of its greatest merits. (*See cut.*) Price, 1st size, large flowering clumps, 35c. each, $3.50 per doz., $20.00 per 100. 2d size, 20c. each, $2.00 per doz., $12.00 per 100.

LARGE-FLOWERING PANSIES.

These are from seed of our own saving, and we can unhesitatingly recommend them either for winter or spring blooming. 50c. per doz.

. . SMILAX VINES. . .

Familiar to every one. Easily grown if trained to strings or stakes. 10c. each, $1.00 per doz.

SWAINSONIA GALEGIFOLIA ALBA.

A plant which is becoming very popular for house culture. Foliage as graceful as an Acacia; flowers pure white, produced in sprays of 12 to 20 flowers each and resembling Sweet Peas. It is of the easiest culture and ever-blooming. 20c. each, $2.00 per doz.

SWAINSONIA GALEGIFOLIA ROSEA.

A fine contrast to the above; rich, rosy red flowers. 20c. each, $2.00 per doz.

HARDY CLIMBING VINES.

READY OCTOBER 25th.

Ampelopsis Veitchii. (Sometimes called "Boston Ivy" and "Japan Ivy.") 1st size, 30c. each, $3.00 per doz.; 2d size, 15c. each, $1.50 per doz.
A. Quinquæfolia. The Old Virginia Creeper.
Akebia Quinata. Dark brown flowers. Fragrant.
Aristolochia Sipho. (Dutchman's Pipe.) 50c. each, $4.50 per doz.
Bignonia Grandiflora. Scarlet Trumpet Vine.
Chinese Matrimony Vine. (*Lycium Chinense.*)
Honeysuckles. Yellow, coral or scarlet, white, pink, evergreen and golden-leaved; fine plants.
Jasminum. Hardy White and hardy Yellow Jessamine.
Wistaria Sinensis. White. Flowers borne in long, drooping clusters. 50c. each.
Wistaria Frutescens. Soft lavender blue. 35c. each.
All of the above **Hardy Climbing Plants** at 25c. each, except where otherwise noted.

HARDY ENGLISH VIOLET

Ready for Delivery about October 25th.

The months named give the period of blooming.

Althea, Double White. Beautiful shrub; double flowers. August and September.

Althea, Double Red. Similar to above except color, which is red. August and September.

Azalea, Hardy. Crimson, pink, yellow, etc. Price, $1.00 each. May and June.

Berberis Thunbergii. One of the handsomest shrubs in cultivation, brilliant red berries in fall and winter.

Calycanthus Floridus. Strawberry scented shrub. June.

Corchorus Japonica. Slender growth; yellow flowers. July to October.

Corchorus Jap. Alba. Similar to the above, but bears pure white flowers; very ornamental. July to October.

Corchorus Jap. Var. Another variety of the preceding, with leaves prettily variegated white and green. July to October.

Cornus Floridus. (Flowering Dogwood.) An interesting species with pure white flowers, followed by showy fruit; symmetrical growth, reaching a height of twelve to thirty feet; excellent for the lawn. June.

Deutzia Candida fl. pl. Double white flowers; a well-known and valuable shrub. June and July.

Deutzia Crenata. Pure white, tinged rose color. June and July.

Deutzia Gracilis. Pure white throughout; low compact habit. June.

Exochorda Grandiflora. Pure white flowers; a grand shrub. May.

Forsythia Viridissima. Yellow flowers; blooms very early. April.

Forsythia Suspensa. (Weeping Forsythia.) A shrub resembling the above in its flowers, but the growth is somewhat drooping. April.

Philadelphus Coronarius. The popular Syringa or Mock Orange. June.

Prunus Pisardi. (Purple-leaved Plum.) Foliage, fruit and shoots of bright purplish red, retaining its color during the heat of summer better than any other purple-leaved tree or shrub. Entirely hardy. Planted with Golden Elder, its beautifully colored foliage presents a most magnificent contrast. May.

Rhodotypus Kerrioides. A very ornamental shrub of medium size, with handsome foliage and large, single, white flowers in the latter part of May, succeeded by numerous small fruits. May.

Sambucus Aurea. (Golden-leaved Elder.) *The finest golden-leaved shrub, and invaluable for producing strong effects in grouping.* Foliage large and handsome, of the richest golden yellow, which it retains throughout the summer, being the most brilliant in color, succeeding best when planted in full exposure to the sun.

Spiræa Callosa. (*Superba.*) Large clusters of pink flowers borne freely. A grand hardy shrub. June to October.

Spiræa Douglasi. A beautiful variety bearing spikes of beautiful rose-colored flowers in July and August.

Spiræa Prunifolia. (Bridal Wreath.) Pure white double flowers completely covering the branches. Crimson foliage in autumn. May and June.

Spiræa Thunbergii. (Thunberg's Spiræa.) Of dwarf habit and rounded graceful form; branches slender and somewhat drooping; foliage narrow and yellowish green; flowers small, white, appearing early in spring, being one of the first Spiræas to flower. Esteemed on account of its neat graceful habit. May.

Spiræa Van Houttei. Flowers pure white, borne in greatest profusion; one of the best in its class. June.

Syringa (Lilac). The well-known purple sweet-scented variety. May.

Syringa Persica Alba. (The White Persian Lilac.) Beautiful sprays of white flowers; slender, graceful growth. A grand plant for cemetery decoration. May.

Viburnum Opulus. (Snowball.) Large drooping white flowers.

Weigelia Candida. Pure white flowers. June to October.

Weigelia Rosea. Rich rose-colored flowers. June and July,

Weigelia Rosea, Var. A variety bearing rose-colored flowers with beautifully variegated foliage. June and July.

All the shrubs on this page we can supply at 25c. each, $2.50 per doz., except where noted.

Japan Maples.

The most ornamental dwarf-growing trees it is possible to imagine. The leaves are fantastically cut and fringed, and the wonderful coloring baffles description, ranging through shades of crimson, scarlet, yellow, and intermediate shades mingled in marvelous harmony. They are entirely hardy. Price, $1.50 each. Ready now.

Japan Snowball.

Foliage olive green through the summer, but toward fall it turns much darker and remains on the plant some time after the first frosts. Flowers pure white, 4 to 6 inches across. Price, large bushes, 2½ to 3 feet, 50c. each, $4.50 per doz. Second size, 12 to 18 inches (mailing size), 30c. each, $3.00 per doz. June.

Hydrangea Paniculata Grandiflora.

The flowers are formed in large white panicles, 9 inches long, which change to a deep pink at the base as the season advances. It grows 5 to 7 feet high and wide, and, as the flowers slightly droop, few plants have the grace and beauty of this shrub. Used largely in cemeteries. Price, strong one-year-old bushes, 1 to 1½ feet high, 25c. each, $2.50 per doz. Extra-strong transplanted bushes, 2½ to 3 feet high and branched, 50c. each, $4.50 per doz. August and September.

Rhododendrons.

These are among the grandest of our hardy flowering shrubs, and cannot be surpassed for lawn decoration. The flowers range through shades of rose, pink, crimson, white, etc., 9 to 12 inches in diameter. Price, plants with 4 to 6 buds, $1.50 each; set of six distinct named sorts for $7.50. Plants with 4 to 8 buds, $1.00 each; set of six sorts for $5.00. May and June. (Ready now.)

Hedge Plants.

Berberry Purple. Very ornamental; purple foliage, scarlet berries. Price, $1.25 per doz., $8.00 per 100.

Pyrus Japonica. (Japan Quince.) Bright scarlet flowers, blooming in early spring. Price, 25c. each, $1.25 per doz., $8.00 per 100.

Privet. (Californian.) A splendid hedge plant; glossy green foliage, white flowers. Price, $1.00 per doz., $6.00 per 100. 2½ feet high.

Note.—We have on hand a grand lot of three year-old shrubs of the varieties named above, and make the following astonishingly low offer:

Collection **A**, of 12 hardy flowering shrubs, 12 distinct varieties					$2.25
" **B**, " 25 "	"	"	25	"	4.00
" **C**, " 50 "	"	"	35	"	7.00
" **D**, " 100 "	"	"	35	"	12.00

Note.—The above collections do not include Rhododendrons, Japan Maples or Azaleas. Prices for larger quantities on application.

CURRANTS ...

If wanted by mail, add 15c. per doz.

(Ready October 25th.)

FAY'S PROLIFIC. This is decidedly the best Red Currant we have. It has been widely planted and has given general satisfaction. The bush is a strong grower, wonderfully prolific, and comes into bearing early. Fruit large, bright red, and of good flavor, and less acid than Cherry, which it is rapidly superseding. $1.50 per doz., $10.00 per 100. *(See cut.)*

LARGE RED CHERRY. The most popular and largest of all red currants except Fay's Prolific. Bunches large, berries very large, bright, sparkling crimson, beautiful, very acid. 2-year, $1.00 per doz., $6.00 per 100.

WHITE GRAPE. The largest and decidedly the best *white* variety, and one of the best of any for the home garden. Bunch large and long; berry large, handsome, translucent white and of best quality, being less acid than others. 2-year, $1.00 per doz., $8.00 per 100.

BLACK NAPLES. A fine *Black* Currant and a general favorite. $1.00 per doz., $6.00 per 100.

Red Raspberries.

(Ready October 25th.)
If wanted by mail, add 10c. per doz.

THOMPSON'S EARLY PROLIFIC. Probably the best early Red Raspberry for general cultivation that we now have. The plant is an excellent grower; canes erect, stout and hardy. Berries are medium to large in size, of a bright crimson color, very productive. It is the earliest red raspberry we have seen, coming into bearing just as the strawberry season is over. Price, 60c. per doz., $3.00 per 100, $20.00 per 1,000.

CUTHBERT. The leading market variety; proved of best general adaptability. Canes hardy and of strong, rampant growth, with large, healthy foliage, and exceedingly productive. Berries large, dark crimson, quite firm and of good flavor. Season late. *(See cut.)*

MARLBORO. The largest of the early red raspberries, ripening a few days later than Hansell. The canes are hardy and fairly productive. Fruit exceedingly large, bright crimson and of fair quality.

TURNER. *(Southern Thornless.)* Extremely hardy and desirable as an early sort for the home garden, but too soft for market purposes. The canes make a strong, healthy growth and are very productive. Berries of good size, bright crimson color, soft and of honeyed sweetness. The plants sucker immediately, and these should be treated as weeds. Early. All the above (except where noted), 50c. per doz., $2.50 per 100, $20.00 per 1,000.

Yellow Raspberry ...

Golden Queen. The most popular and best yellow Raspberry yet introduced. Of large size, great beauty, high quality, hardiness and productiveness. Fully equal to Cuthbert in size of fruit and vigor of growth. Price, 50c. per doz., $2.50 per 100, $20.00 per 1,000.

LOVETT'S BEST BLACKBERRY.

We strongly recommend **Fall planting** for all small fruits, if they can be put in the ground before it is frozen up for the winter. If they cannot be planted, then we would still advise our customers in the Northern States to procure their supply in the fall, and "heel them in" in some sheltered situation, so that they may plant them early in the spring, just as soon as it is free from frost, and dry enough to cultivate. This advice, if followed, will be of practical benefit to many who have hitherto failed or have been only partially successful.

BLACKBERRIES

(Ready October 25th.)

HOW THEY ARE GROWN. The instructions which we give for Blackberries apply also to Raspberries, both red and black. The Blackberries we offer are all strong **Root Cutting** plants, which are vastly superior to the "sucker plants" so largely sold. The varieties offered are the very best on the market. For garden culture Blackberries should be planted in rows five feet apart with three feet between the plants in the rows. Thin out by cutting away all the canes which have borne fruit; after the crop is gathered clean up the ground and cultivate between the rows and the plants. When the new shoots are about four feet high pinch out the tips to stop them; this causes an extra growth of "laterals" or side shoots, and when these are twelve to eighteen inches long the tips should be pinched out of them also. The result of this work will be to form a stocky plant and insure a good crop of good fruit, in sharp contrast to the scrambling plants usually seen with a small crop of poor quality fruit.

LOVETT'S BEST. A thoroughly reliable Blackberry of large size, with a cane of ironclad hardiness. Unites not only these two invaluable properties in an eminent degree, but possesses in addition the merits of ripening early, great productiveness, entire freedom from disease and double or rose blossom, strong, vigorous growth of cane, extra high quality, jet-black, permanent color and fine appearance. *(See cut.)* Price, 60c. per doz., $4.00 per 100, $30.00 per 1,000.

Wilson Junior. Takes the place of the old Wilson's Early; it possesses all its good qualities and is hardier and more productive, combining size, earliness and productiveness with the fine appearance and market properties of that variety.

ERIE. The most popular of all the standard Blackberries. The canes are of ironclad hardiness, of the strongest growth, free from rust, double blossom and all other diseases, and wonderfully productive, bending the robust cane to the ground with the weight of fruit. The berry is of the very largest size, of excellent quality.

Early Harvest. The earliest Blackberry except Early King. The berries are not of the largest size, but very uniform and of a bright glossy blackness that renders them extremely enticing. For the South its value is sometimes over-estimated, and its early ripening brings it into market at a time when it has no competitors.

Kittatinny. Once the most popular of all Blackberries for general planting and very fine for main crop. The berries are large, handsome and of delicious flavor; canes of strong, erect growth and productive. Season medium to late.

Snyder. Valuable for the North by reason of its extreme hardiness. Wonderfully productive, small to medium in size, sweet, juicy flavor.

DEWBERRY, Lucretia (or Creeping Blackberry), conceded to be the finest of its class, as early as Early Harvest and as large as the Erie Blackberry. The quality is superb, juicy and melting. Set the plants in rows six feet apart, and three feet between the plants in the rows. Keep the soil mellow and clean at all times. Price, for any of the above (except where noted), 50c. per doz., $2.50 per 100, $20.00 per 1,000. *If wanted by mail, postage must be added at the rate of 10c. per doz.*

CUTHBERT RASPBERRY.

MOORE'S DIAMOND GRAPE.

STRAWBERRY PLANTS. *POT GROWN.*

Space will not permit us to give descriptions of the numerous varieties which we offer, all of which are fully described in our Strawberry Catalogue, a copy of which may be had upon application by any who did not receive it.
Ground Layers we can supply, if wanted, after October 25th, at half the rates quoted in that catalogue.

GOOSEBERRIES. *(Ready October 25th.)*

Downing. A vigorous variety, not much affected by mildew. 75c. per doz., $5.00 per 100.
Industry. New English variety, the finest of all Gooseberries. 15c. each, $1.50 per doz., $10.00 per 100.

BLACK RASPBERRIES. *"CAPS."*

(Ready October 20th. If wanted by mail, add 10c. per doz.)

Lovett Raspberry. The Lovett Raspberry is of ironclad hardiness, and is the strongest in growth of cane of any. In the bone garden especially, its sweet, fine flavor and small seeds make it welcome. Very profitable for market. 60c. per doz., $4.00 per 100, $30.00 per 1,000.
Progress. (*Pioneer.*) Is a most profitable market sort, entirely hardy. Berries jet black, very firm and of good quality.
Palmer. A new variety and an improvement on **Souhegan**, from which it sprung. Very early and a grand fruit for either family or market, vigorous and hardy, with foliage healthy and free from rust; wonderfully productive.
Ohio. Exceedingly productive, very hardy and free from disease; berries of good size, jet-black and of excellent quality.
Gregg. A popular market sort. Canes of strong, vigorous growth; berries very large, covered with heavy bloom, firm, meaty, and of fine flavor. It responds liberally to generous treatment.
Shaffer's Colossal. Canes are of wonderful vigor and size, hardy and enormously productive. Berries large, rather soft, but luscious, and of a rich, sprightly flavor.
Souhegan, or Tyler. A very early Blackcap, and the leading early market sort. Canes vigorous and hardy, wonderfully productive. Fruit of good size, jet-black, with but little bloom; firm, sweet and pleasant.
All the above (except where noted), 50c. per doz., $2.50 per 100, $15.00 per 1,000.

HARDY GRAPES IN VARIETY. *(Ready October 25th.)*

Agawam. (Rogers' No. 15.) Berries large, of *bronze* color; bunches of good size and form. 25c. each, $2.50 per doz.
Brighton. This still remains the best grape in its color for table use. In color, form and size of both bunch and berry, it resembles Catawba, but ripens early—with the Delaware. Vine a free grower and productive. Price, 2-year vines, 25c. each, $2.50 per doz.
Concord. One of the best of sorts. Bunch and berries large; color *black* with a rich bloom. 20c. each, $2.00 per doz.
Delaware. *Red*; bunches compact, berries small, sweet, and of the most excellent flavor. 25c. each, $2.50 per doz.
Eaton. This promising new Grape is similar in foliage to Concord, *and in growth, health, hardiness and quality is in every respect its equal*, while in size of bunch and berry it is much larger and more attractive in appearance. Pulp tender, separating freely from the seeds and dissolving easily in the mouth. Very juicy, ripens with Concord or a little earlier. Price, 2-year vines, 30c. each, $3.00 per doz.
Goethe. (Rogers' No. 1.) Bunches medium to large, occasionally shouldered; berries very large, oblong, of a yellowish green, sometimes blotched, with a pale red toward the sun and entirely red when fully ripe. A sweet, vinous, juicy grape, with a peculiar delicious aroma. 25c. each, $2.50 per doz.
Golden Pocklington. Very large, *deep amber* in color. One of the most attractive grapes grown. 25c. each, $2.50 per doz.
Lindley. (Rogers' No. 6.) Bunch medium, somewhat loose; berry medium to large, round; color, *a rich shade of red*, rendering it a very handsome and attractive grape; flesh tender, sweet, with an aromatic flavor; ripens soon after the Delaware. 25c. each, $2.25 per doz.
Moore's Diamond. In vigor of growth, color and texture of foliage, with hardiness of vine, it is the equal of its parent "Concord," while in quality the fruit is equal to many of our hothouse grapes. It is amongst the earliest and ripens from two to four weeks ahead of "Concord." (*See cut.*) 30c. each, $3.00 per doz.
Moore's Early. Resembling the Concord in style of growth and berry; ripening two weeks earlier. 30c. each, $3.00 per doz.
Martha. One of the best *greenish white* grapes; exquisite flavor. 20c. each, $2.00 per doz.
Merrimack. (Rogers' No. 19.) Large berry; *jet-black*; fine quality. 25c. each, $2.50 per doz.
Niagara. A grand *white* grape, hardy, fine quality. 25c. each, $2.50 per doz.
Salem. (Rogers' No. 53.) A splendid *coppery red* grape ripening with Concord, tender, juicy; one of the very best. 25c. each, $2.50 per doz.
Worden. *Black.* Very juicy, large size, early. 25c. each, $2.50 per doz.
Wilder. (Rogers' No. 4.) An excellent variety. Berries medium; bunches large; color *black.* A good bearer. 25c. each, $2.50 per doz.

SPECIAL OFFER.

Full set of 16 Hardy Grapes for $3.50. (If wanted by mail, add 25c. per set.)

FOREIGN GRAPE VINES.

Black Hamburg, Muscat of Alexandria, Bowood, Muscat, Gros, Colmar. Strong one-year vines. $1.00 each, $9.00 per doz.

RHUBARB ROOTS. *(Ready October 25th.)*

St. Martin's. A new English variety, now grown largely for the London market. It is not only immensely productive, but it is also one of the earliest, and, above all, it has a rich, spicy flavor, very similar to the gooseberry, when used for pies or tarts. 20c. each, $2.00 per doz.
Linnæus. The standard variety. 15c. each, $1.50 per doz.

"LOVETT" BLACK RASPBERRY.

A FEW NOVELTIES AND SPECIALTIES IN VEGETABLE SEEDS

WE DELIVER FREE

to any Post Office or Railroad Express Offfice
in the UNITED STATES,
at prices quoted in this Catalogue, all
BULBS, VEGETABLE AND
FLOWER SEEDS,
except where noted.

HENDERSON'S PALMETTO ASPARAGUS ROOTS.

(Ready in November for the South, and in March and April for the North.)

Early, large, tender, delicious. The finest Asparagus grown. The Palmetto is not only much earlier, but is also a better yielder, and is more even and regular in its growth than any other variety, and must eventually supplant all older sorts. Average bunches, containing fifteen shoots, measure 11½ inches in circumference, and weigh about two pounds. The Palmetto has now been planted in all parts of the country, and it has been proven equally well adapted for all sections North and South. Its quality is unequaled. *(See cut.)*

Palmetto. Asparagus roots (a saving of one to two years is effected by planting roots.) Splendid two year-old. $1.50 per 100, $10.00 per 1,000. 50 roots at 100 rate ; 500 at 1,000 rate. If to be sent by mail, add 40c. per 100 to the prices.

HENDERSON'S "EARLY SNOWBALL" CAULIFLOWER.

FAMOUS THE WORLD OVER.

BEYOND QUESTION
THE BEST
CAULIFLOWER

For either Early or Late,
For Family or Market Garden,
For Forcing or Cold Frames.

...IT HEADS where others fail.

Those who have never succeeded before should grow

"HENDERSON'S SNOWBALL."

It is hardly necessary for us to describe this variety at all, as its name has now become a household word wherever this delicious vegetable is grown. **Snowball Cauliflower is the standard everywhere for quality** with the seedsman, the market gardener and the amateur. While originally recommended and used as an early variety, it has not only supplanted all other sorts for early spring planting, but it has largely driven out the large late sorts for fall use, being much finer in quality, and the only Cauliflower that is absolutely certain to head when the conditions are right, forming a perfect snow-white head, averaging nine inches in diameter.

HENDERSON'S EARLY SNOWBALL CAULIFLOWER is superior to all others. It is the earliest of all Cauliflowers. Its close-growing, compact habit enables one third more to be planted on the same space of ground than can be done with any other variety. For forcing under glass during winter and spring, this **Early Snowball** variety is peculiarly well adapted, from its dwarf growth and short outer leaves, and for this purpose no other Cauliflower is now so largely grown. *(See cut.)* Price, 25c. per pkt., $1.00 per oz., $14.00 per ¼ lb., $48.00 per lb.

HENDERSON'S EARLY
SNOWBALL CAULIFLOWER.

... HENDERSON'S ...
"EARLY SPRING" CABBAGE.

A NEW EXTRA EARLY CABBAGE, WITH A ROUND FLAT HEAD, COMING IN WITH THE WAKEFIELD.

The most valuable early Cabbage ever introduced. The variety is of Early Summer type, but about one-fifth smaller, having only four or five outside leaves, and these so small, and growing so near to the head, that it may be successfully planted twenty-one inches apart, as close as any variety we know of. But **its great value lies in the fact that it is the only first Early Flat Cabbage.** It possesses wonderful uniformity in shape, being round, slightly flattened at top of head. Stem is short and extends but little into the head; this feature is valuable, as it makes almost the entire head edible. The veining of the leaves is particularly fine, in which respect it shows its finely bred character, being entirely free from any coarseness whatever. There is no Cabbage we know of having a more solid head; but, added to this, it has the peculiarity of heading firmly at an early stage in its growth, so that the finest Cabbage, though small, can be obtained long before it has attained its mature size. It is entirely free from any rankness of flavor. (*See cut.*) Price, 20c. per pkt., $1.00 per oz., $3.00 per ¼ lb.

HENDERSON'S NEW "EARLY SPRING" CABBAGE.

HENDERSON'S EARLY JERSEY WAKEFIELD CABBAGE.

. . HENDERSON'S . .
Selected Early Jersey Wakefield.

ORIGINAL STOCK.   **THE STANDARD EARLY CABBAGE.**

The merits and characteristics of the **Early Jersey Wakefield Cabbage** are now so well known as to hardly need repeating here; still as our catalogue annually falls into the hands of thousands who have not before seen it, we may state that it is universally considered the best early Cabbage in cultivation. Among its merits may be mentioned its large size of head *for an early sort*, small outside foliage, and its uniformity in producing a crop. The heads are pyramidal in shape, having a blunted or rounded peak. Our seed of this famous Cabbage is very choice, being saved from selected fine hard heads. (*See cut.*) Price, 5c. per pkt., 30c. per oz., $1.00 per ¼ lb., $3.00 per lb.

"Had Peter Henderson done nothing else but introduce the Early Wakefield Cabbage, it would be a lasting monument to his name. No better variety is in cultivation to-day among the early kinds."—*Country Gentleman.*

THE CHARLESTON OR LARGE TYPE OF WAKEFIELD.

For some years past customers have requested us to procure for them, if possible, a Cabbage which had all the characteristics of the Early Jersey Wakefield, but of a greater size. With this in view, we carefully selected from one of our best stocks of Wakefield a larger type, which is now so fixed in its character that we can offer it with confidence to those desiring a Cabbage of this kind. This selection will average about 50 per cent. larger in size than the old type of Wakefield, and is only two or three days later. Price, 10c. per pkt., 40c. per oz., $1.25 per ¼ lb., $4.00 per lb.

Henderson's Succession Cabbage. .
IS POSITIVELY THE
FINEST CABBAGE IN EXISTENCE,
Whether for Medium Early, Main Crop or Winter use. PERFECT in every respect.

The Succession Cabbage we consider one of our most valuable contributions to horticulture. It would be classed as a second early, coming in a few days later than Early Summer, but it is immeasurably superior to that variety; it is of nearly double the size and is absolutely true to its type under all conditions. In addition to this, it has no tendency whatever to run to seed. We can say without exaggeration that *it is the finest Cabbage in existence* to-day; whether for medium early, main crop or late use it has no equal. It is so finely bred and so true to type that in a field of twenty acres every head appears alike. We can recommend it either for the market gardener, trucker or private planter, as it is a perfect cabbage in every respect, not only being of the largest size, but of handsome color and of the finest quality. It is probably the safest variety for an amateur to plant as it does well at all seasons, and one is almost sure of getting a crop, no matter when it is planted. Our stocks of Cabbage of all varieties have for years been the *acknowledged standard of excellence in this country*, and when we state that we consider **Succession** to be the most valuable variety that we have ever introduced, our opinion of its great merit will be apparent to all. (*See cut.*)
Price, 10c. per pkt., 40c. per oz., $1.25 per ¼ lb., $3.50 per lb.

HENDERSON'S SUCCESSION CABBAGE.

PRICE LIST OF
VEGETABLE SEEDS
For Fall Sowing

WE DELIVER FREE To any Post Office or R. R. Express Office in the United States, at prices quoted in this Catalogue, all Bulbs, Vegetable and Flower Seeds, except where noted.

ASPARAGUS ROOTS.

(Ready in November for the South, in March and April for the North.)

If to be sent by mail, add 40 cts. per 100 to the prices.

The Palmetto. Splendid roots. $1.50 per 100, $10.00 per 1,000. (*50 roots at 100 rate; 500 at 1,000 rate.*)

Colossal. Fine two-year-old roots at $1.00 per 100, $6.00 per 1,000. (*50 roots at 100 rate; 500 at 1,000 rate.*)

BEANS, Dwarf Green Podded.

If Beans are desired by mail, please add 5 cts. per pint and 10 cts. per quart for postage.

	pkt.	pt.	qt.	pk.	bu.
China, Early	10	20	30	1.15	4.00
Dwarf Case Knife	10	30	35	1.75	5.50
Dwarf Horticultural	10	25	35	1.35	4.50
Longfellow	10	20	35	2.00	7.00
Mohawk, Early	10	15	25	1.15	4.00
Refugee, Extra Early	10	20	30	1.35	4.50
" or 1,000 to 1	10	15	25	1.00	3.50
Six Weeks Long Yellow	10	15	25	1.15	4.00
Valentine, Earliest Red	10	20	30	1.35	4.50
" Early Red	10	15	25	1.15	4.00
" White	10	20	35	1.35	4.50
Warwick, Early	10	20	35	2.00	7.00

BEANS, Dwarf Wax or Yellow Podded.

	pkt.	pt.	qt.	pk.	bu.
Black Wax, or Butter	10	20	30	1.25	4.00
Cylinder Black Wax	10	20	35	1.30	5.00
Flageolet Wax	10	20	35	1.60	5.50
Golden Eyed Wax	10	20	35	1.50	5.00
Golden Wax	10	20	30	1.25	4.00
" Improved	10	20	30	1.35	4.50
" Keeney's Rustless	10	20	40	1.75	6.00
Refugee Wax	10	20	35	1.50	5.00
Valentine Wax	10	20	40	1.75	6.00
Wardwell's Kidney Wax	10	20	30	1.35	5.00
White Seeded Wax	10	20	30	1.60	5.50
Yosemite Mammoth Wax	10	20	35	2.50	9.00

BEANS, Dwarf Lima.

	pkt.	½pt.	pt.	qt.	pk.
Burpee's Bush Lima	10	20	30	50	3.00
Dreer's Bush Lima (*Kumerle or Thorburn's*)	10	15	25	45	3.00
Henderson's Bush Lima	10	15	25	40	2.50

BEET.

	pkt.	oz.	¼lb.	lb.
Arlington Improved	10	15	25	70
Bassano, Early Flat	5	10	20	55
Bastian's Blood Turnip	5	10	20	55
Blood Turnip, Early	5	10	20	55
Dewing's Improved	5	10	20	60
Eclipse	5	10	20	60
Edmund's	5	10	20	60
Egyptian Turnip	5	10	20	50
Half Long	10	15	35	1.00
Electric	10	15	35	1.00
Lentz	5	10	20	65
Long Smooth Blood	5	10	20	55
Swiss Chard	5	10	20	60
Yellow Turnip	5	10	20	60

BRUSSELS SPROUTS.

	pkt.	oz.	¼lb.	lb.
Dalkeith	10	30	1.00	
Dwarf Improved	10	25	80	3.00
Tall French	5	20	70	2.50

CABBAGE, First Early.

	pkt.	oz.	¼lb.	lb.
Early Spring (*see cut and description, page 65*)	20	1.00	3.00	
Early Jersey Wakefield (*see cut and description, page 65*)	5	30	1.00	3.00
Charleston Wakefield (*see description, page 65*)	10	40	1.25	4.00
Early York	5	20	50	2.00
Early French Oxheart	5	20	60	2.00

CABBAGE, Second Early.

	pkt.	oz.	¼lb.	lb.
Early Summer (*Henderson's*)	5	25	80	3.00
Early Winningstadt	5	20	50	2.00
Fottler's Imp'd Brunswick	5	20	50	2.00
Large Early York	5	20	60	2.00
Succession (Henderson's) (*see cut and description, page 65*)	10	40	1.25	3.50

CABBAGE, Late or Winter.

	pkt.	oz.	¼lb.	lb.
Autumn King	10	40	1.25	3.50
Premium Flat Dutch	5	20	60	1.75
Selected Flat Dutch (*Henderson's*)	10	20	70	2.50

CABBAGE, Red.

	pkt.	oz.	¼lb.	lb.
Mammoth Rock Red	10	40	1.25	4.00
Red Dutch	5	25	75	2.50

CABBAGE, Savoy.

	pkt.	oz.	¼lb.	lb.
American Drumhead	10	25	75	2.50
Netted Savoy	5	25	75	2.50

CARROT.

(*Series marked * are the best to grow for stock feeding.*)

	pkt.	oz.	¼lb.	lb.
Carentan Early Half Long, Scarlet	5	10	30	90
Chantenay	5	10	30	90
Danvers*	5	10	30	90
French Forcing	5	15	40	1.25
Early Half Long, Scarlet, Pointed	5	10	30	90
Half Long, Red, Stump-rooted	5	10	30	90
Intermediate	10	15	30	90
Long Orange Improved*	5	10	30	80
New York Market	10	15	50	1.50
Oxheart	10	15	30	1.00

CAULIFLOWER.

	pkt.	oz.	¼lb.	lb.
Algiers, Large Late	10	80	2.75	
Early London	5	75	2.50	
Extra Early Dwarf Erfurt	15	3.00	10.00	
Extra Early Paris	5	80	2.75	
Half Early Paris	5	75	2.50	
Henderson's Snowball (*see cut and description, page 54*)	25	4.00	14.00	48.00
Lenormand's Short Stem	5	75	2.50	
Veitch's Autumn Giant	5	60	1.75	

CORN SALAD.

	pkt.	oz.	¼lb.	lb.
Large Leaved	5	10	25	80

CUCUMBER.

	pkt.	oz.	¼lb.	lb.
Boston Pickling	5	10	25	70
Cool and Crisp	10	20	50	1.50
Early Cluster	5	10	25	70
Early Frame	5	10	25	70
Early Russian	5	10	25	70
Everbearing	5	10	25	80
Evergreen (*Livingston's*)	5	10	25	80
Extra Early Green Prolific	10	15	30	80
Green Prolific	5	10	25	70
Japanese Climbing	10	25	50	2.00
Jersey Pickling	5	10	25	70
Long Green	5	10	25	75
Long Green Turkey	5	10	25	80
Nichol's Medium Green	5	10	25	70
Short Green	5	10	25	70
Tailby's Hybrid	5	10	25	70
West India Gherkin	5	15	35	1.25
White Spine, Imp'd Early	5	10	20	70
" " Extra Long	5	10	25	70

English Frame Varieties.

Telegraph, Lord Kenyon's Favorite, Duke of Edinburgh, Cuthill's Black Spine, 25c. pkt. each.

EGG PLANT.

	pkt.	oz.	¼lb.	lb.
Black Pekin	10	50	1.50	
Early Long Purple	5	30	80	2.50
New York Improved	5	40	1.25	4.50
" " " Spineless	10	50	1.50	

KALE.

	pkt.	oz.	¼lb.	lb.
Brown German Curled	5	15	35	1.00
Dwarf Green Curled Scotch	5	15	35	1.00
Siberian	5	10	20	60

Lettuce, Head Varieties.

	pkt.	oz.	¼lb.	lb.
All the Year Round	5	15	40	1.25
Big Boston	10	25	70	2.00
Black Seeded Butter	5	15	40	1.25
Boston Market	5	15	40	1.25
Deacon	5	15	40	1.25
Drumhead, or Malta	5	15	40	1.25
Golden Queen	10	25	75	2.50
Hanson	5	15	40	1.25
Hardy Green Winter	5	15	40	1.25
Large White Summer	5	15	40	1.25
Mignonette	10	40	1.25	4.50
New York, Henderson's	10	25	65	2.00
Salamander Perfected	10	20	50	2.00
Tennis Ball	5	15	40	1.25
Yellow Seeded Butter	5	15	40	1.25

LETTUCE, Curled or Loose-leaved Varieties.

	pkt.	oz.	¼lb.	lb.
Boston Curled	5	15	40	1.25
Defiance Summer	5	15	40	1.25
Early Prizehead	5	15	40	1.25
Large India	5	15	40	1.25
Simpson, Early Curled	5	15	40	1.25
" Black Seeded	5	15	40	1.25

GOLDEN QUEEN LETTUCE.   AMERICAN WONDER PEA.   FRENCH BREAKFAST RADISH.   THICK-LEAVED SPINACH.

COS LETTUCE.

	pkt.	oz.	¼lb.	lb.
Paris White	5	20	50	1.50
Trianon	10	25	75	2.50

MELON, Musk.

	pkt.	oz.	¼lb.	lb.
Baltimore, or Acme	5	10	25	80
Banquet	10	20	50	1.50
Delmonico, Perfected	10	25	75	2.00
Emerald Gem	5	10	30	90
Golden Netted Gem	10	15	30	1.00
Hackensack	5	10	25	80
" Early	10	20	40	1.50
Improved Christiania	5	10	30	1.00
Jenny Lind	5	10	25	80
Large Yellow Cantaloupe	5	10	25	80
Miller's Cream or Osage	5	10	25	80
Montreal Market	5	10	25	80
Newport	10	35	1.00	3.00
Nutmeg	5	10	25	80
Skillman's Netted	5	10	25	80

MUSHROOM SPAWN.

English, 15c. lb.; 8 lbs. for $1.00; by mail, 25c. lb.
French, 2-lb. box, 75c.; by mail, 95c.

ONION.

	pkt.	oz.	¼lb.	lb.
Extra Early Flat Red (5 lbs., $7.00)	5	15	50	1.50
Wethersfield Large Red (5 lbs., $5.50)	5	15	40	1.20
White Portugal, or Silver Skin	5	25	80	2.50
Round Yellow Danvers (5 lbs., $4.50)	5	15	40	1.00
Yellow Dutch 5 lbs., $4.50)	5	15	40	1.00

ONION, Globe-shaped Varieties.

	pkt.	oz.	¼lb.	lb.
Southport White Globe	10	25	80	2.50
" Yellow Globe	10	20	40	1.20
" Red Globe	5	15	40	1.20
Yellow Globe Danvers	5	15	40	1.10

ONION, Italian and Spanish Varieties.

	pkt.	oz.	¼lb.	lb.
Adriatic Barletta	10	25	70	2.00
Giant Rocca	5	15	40	1.60
Giant White Garganus Silver King	10	20	50	1.50
Mammoth Pompeii (Red Garganus)	10	20	50	1.50
Neapolitan Marzajola	5	20	60	2.00
Queen	5	20	60	2.00
Prizetaker	10	25	60	2.00
Tripoli Large White Italian	5	20	50	1.50
" Large Red Italian	5	20	50	1.50
White Bunch	5	20	60	2.00

ONION, Bermuda Varieties.

	pkt.	oz.	¼lb.	lb.
Pale Red	5	30	1.20	4.00
White	5	35	1.20	4.00

ONION SETS.

If wanted by mail, add 10c. per quart for postage.

	qt.	pk.
Red	.30	2.00
White	.35	2.50
Yellow	.30	2.00
Potato	.35	2.50
Top, or Button	.30	2.00
Shallots	.30	2.00

PARSLEY.

	pkt.	oz.	¼lb.	lb.
Champion Moss Curled	5	10	30	1.00
Emerald	10	15	30	1.00
Extra Double Curled	5	10	30	90
Fern Leaved	5	10	30	1.00
Hamburg Turnip Rooted	5	10	30	1.00
Plain	5	10	25	80

PEAS, Early.

If Peas are desired by mail, please add 5c. per pint and 10c. per quart for postage.

	pkt.	pt.	qt.	pk.	bu.
Admiral	10	20	30	1.50	5.00
Alaska	10	15	25	1.25	4.00
Alpha, Laxton's	10	20	30	1.50	5.00
American Wonder	10	20	30	1.70	5.00
Blue Beauty	10	20	30	1.25	4.50
Chelsea	15	25	40	2.00	7.00
Daniel O'Rourke	10	15	25	1.00	3.50
First of All	10	20	30	1.25	4.00
Little Gem	10	15	25	1.25	4.50
Nott's Excelsior	10	20	35	2.00	7.00
Premium Gem	10	20	30	1.50	4.75
Tom Thumb	10	20	30	1.50	5.00

PEAS, Second Early and Medium.

	pkt.	pt.	qt.	pk.	bu.
Abundance	10	15	25	1.30	4.50
Advancer	10	15	25	1.50	4.50
Fillbasket	10	15	25	1.50	5.00
Heroine	10	25	40	1.75	5.50
Horsford's Market Garden	10	15	25	1.25	4.00
Shropshire Hero	10	20	30	1.50	5.00

PEAS, Main Crop and Late.

	pkt.	pt.	qt.	pk.	bu.
American Champion	15	25	40	2.50	8.50
Black-eyed Marrowfat	10	20	75	2.50	
Champion of England	10	15	25	1.50	4.50
Everbearing	10	20	30	1.50	5.00
Juno	15	25	45	2.00	7.00
Pride of the Market	10	20	30	2.00	7.00
Queen	10	20	40	1.50	5.00
Stratagem	10	20	35	1.60	5.50
White Marrowfat	10	20	75	2.50	

RADISH, Round Varieties.

	pkt.	oz.	¼lb.	lb.
Early Scarlet Turnip	5	10	20	60
Early Scarlet Turnip, White Tipped	5	10	20	65
Early White Turnip	5	10	20	65
Newcom White	5	10	25	75
Early Rose Turnip	10	20	50	1.50
Early Round Dark Red	5	10	20	65
Early Scarlet Globe	10	15	40	1.00
Rapid Forcing	5	15	40	1.00
Red Forcing	5	10	30	80
Yellow Summer Turnip	5	10	20	70

RADISH, Olive-shaped.

	pkt.	oz.	¼lb.	lb.
French Breakfast	5	10	25	80
Red Rocket	5	15	30	1.00
Olive-shaped Scarlet	5	10	20	65
" White	5	10	20	65
Oval-shaped Yellow	10	20	50	1.50

RADISH, Long Varieties.

	pkt.	oz.	¼lb.	lb.
Brightest Scarlet (Cardinal)	10	15	30	80
Celestial	10	15	40	1.25
Chartier (Beckert's)	5	10	25	75
" New White	10	20	50	1.50
Giant White Stuttgart	5	10	25	80
Long Scarlet, Short Top	5	10	20	65
Long White Vienna Lady Finger	5	10	30	90
White Strasburg	5	10	25	80
Wood's Early Frame	5	10	20	65

RADISH, Winter Varieties.

	pkt.	oz.	¼lb.	lb.
California Mammoth, White	5	10	15	85
Long Black Spanish	5	10	25	80
Rose China, Winter	5	10	30	1.00
Sandwich	10	25	70	

SPINACH.

	pkt.	oz.	¼lb.	lb.
Large Round Viroflay	5	10	15	35
Long Standing	5	10	15	35
New Victoria	5	10	15	40
Prickly	5	10	15	40
Round Leaved (Bloomsdale)	5	10	15	35
Savoy Leaved Norfolk Thick Leaved	5	10	15	40

TOMATO.

	pkt.	oz.	¼lb.	lb.
Acme	5	20	60	2.00
Beauty	5	20	70	2.50
Crimson Cushion	15	60	2.00	
Dwarf Champion	5	25	75	2.50
Early Ruby	10	30	80	2.50
Essex Hybrid	5	20	60	2.00
Favorite	5	20	60	2.00
Golden Sunrise	5	25	70	2.50
Ignotum	5	20	70	2.50
Lorillard	5	20	70	2.50
Mikado	10	25	70	2.50
Paragon	5	20	70	2.50
Peach	10	25	80	3.00
Pear-shaped, Red	5	20	70	2.50
Perfection	5	20	70	2.50
Ponderosa	10	60	2.00	
Red Cherry	5	20	70	2.50
Strawberry, or Winter Cherry	5	20	70	2.50
Table Queen	10	30	1.00	3.00
Trophy	5	20	70	2.50
Trophy, extra selected	10	30	1.00	3.00
Yellow Plum	5	20	70	2.50

TURNIP.

	pkt.	oz.	¼lb.	lb.
Early White Milan	10	20	60	2.00
Aberdeen, Yellow Purple Top	5	10	20	60
Early Flat Dutch	5	10	20	55
Extra Early Milan, Purple Top	10	20	60	1.50
Golden Ball	5	10	25	75
Long White Cowhorn	5	10	20	60
Purple Top White Globe	5	10	20	60
Red Top Strap Leaf	5	10	20	55
Scarlet Kashmir	10	20	60	2.00
Seven Top (for Greens)	5	10	20	55
Snowball	5	10	25	75
White Egg	5	10	20	60
White Strap Leaf	5	10	20	55
Yellow Globe	5	10	25	70
Yellow Stone	5	10	25	75

HENDERSON'S Unrivaled Pansies For Sowing IN Autumn

Henderson's Mammoth Butterfly Pansies.

A strain of Pansy which for variety and beauty has never been excelled. Of matchless forms, colors and markings, *with flowers half as large again* as ordinary Pansies. They will both astonish and delight "Pansy Fanciers." **Mixed Colors.** Per packet, 30c.
Collection of 12 *types, as shown on the colored plate in our manual of* "*Everything for the Garden,*" 1897, $1.25.

	Per Pkt.
Good, mixed	5
German, finest mixed	10
New Fancy Belgian, mixed	15
English Prize-taker Show, mixed	35
English Prize-taker Fancy, mixed	50
Cassier's 5 Blotched Odier, mixed	25
Cassier's 5 Blotched Odier, on white	25
Cassier's Odier, on yellow	25
" " on red	25
Bugnot's Large Stained Parisian, mixed	50
All Nations, mixed	25
Giant Trimardean, mixed colors	15
" " violet and blue	15
" " marbled yellow	15
" " white	15
" " purple	15
" " black	15
" " lavender and violet	15
" " blue, blotched black	15
" " snow-white	25
" " white with violet eyes	15
" " yellow, mahogany eyes	15
" " yellow, margined blue	15
" " striped	15
" " pink	15
" " Fire King	15
" " collection of 14	$1.50

	Per Pkt.
Azure Blue, lavender, shaded dark blue	5
Black with Violet, black zoned violet	10
Blood Red (Victoria), velvety blood red	15
Bronze, golden bronze	15
Coquette de Poissy, reddish lavender	10
Emperor William, deep blue, black blotches	10
Fire Dragon, yellow, claret and maroon	10
Garnet Red, gold banded	10
" silver banded, garnet margined	10
Golden Brown, gold and cinnamon	10
Gold Else, pure bright yellow, profuse	10
Gloriosa Perfecta, blue, cream, garnet and yellow	15
King of the Blacks (Faust), jet black	5
Lord Beaconsfield, lavender and violet	5
Mahogany, rich mahogany brown	10
" banded pink, yellow and white	10
Marbled Silver Claret and Purple	10
" " Gold	10
Pres. Carnot, white, five large violet blotches	15
Rose Marbled, rosy lilac, veined claret	10
Rex, solid deep purple	5
Striped, chocolate, white, lilac and claret	5
Snow Queen, pure snow-white	5
Violet, banded silver, violet, margined white	10
White Violet Eyes (White Treasure)	5
Yellow with Dark Eyes	5
Collection of German, 6 sorts	25
" " 12 "	50

Pansies sown in the fall produce Incomparably the finest and largest flowers of the brightest colors.

A FEW FLOWER SEEDS SUITABLE FOR AUTUMN SOWING.

Varieties requiring the protection of Cold Frames in the cold latitudes.

	Per Pkt.
Antirrhinum, tall, mixed	5
" choice striped, mixed	10
" Tom Thumb, "	5
Auricula, finest mixed	10
Bellis, double, mixed	10
" white	10
" rose	10
" Giant Snowball	15
" Red	15
Canterbury Bells, single, mixed	5
" double,	5
" Calycanthema "	5
Cowslips, mixed	5
Hollyhock, double, mixed	10
" extra choice, mixed	10
" 12 separate colors, each	10
" collection, 12 sorts	1.00
" 6 "	50
Myosotis Alpestris, mixed	5
" white	5
" blue	5
" Victoria, "	10
" Palustris, "	10
Pansies. (See top of page.)	
Pentstemon, mixed	10
Polyanthus, "	5
Primula Cortusoides	10
" Sieboldi	25
" Japonica	10
" Rosea	25
Pyrethrum Aureum	5
" Selaginoides	10
Rose, Tea Scented, double	10
" Little Midget	20
Viola Odorata, mixed	10

Varieties requiring House or Conservatory Culture.

	Per Pkt.
Abutilon, mixed colors	15
Aralia Sieboldi	10
Asparagus Plumosus Nanus	25
Begonia, Tuberous, single, mixed	25
" double, "	50
" Vernon	10
Calceolaria, large-flowering, mixed	25
Carnation, Marguerite, Giant,	15
" finest, double, mixed	15
Centaurea Candidissima	10
" Gymnocarpa	10
Cineraria, large-flowering, mixed	25
Cobæa Scandens, blue	10
Coleus, mixed	15
Cyclamen Persicum, mixed	10
" Giganteum	25
Ferns, mixed	10
Fuchsia, "	10
Geranium, "	10
Gloxinia, large-flowering, mixed	25
Grevillea Robusta	10
Heliotrope, mixed	5
Lantana, "	10
Lobelia Compacta, mixed	10
Maurandia, mixed	10
Mignonette, Machet	10
" Hybrid Spiral	5
Musa Ensete	25
Petunia, striped, mixed	10
" Giant of Calif., mixed	25
Primrose, large-flowering, mixed	25
" white	25
Rhodochiton	15
Smilax	10
Torenia Fournieri	10

Varieties that can be sown in the open ground to come up next spring.

	Per Pkt.
Alyssum, Sweet	5
" Tom Thumb	5
" Saxatile Compacta	5
" Wiersbecki	5
Aquilegia, single, mixed	5
" double, "	5
Calliopsis, mixed	5
" Golden Wave	5
" Tom Thumb, mixed	5
Campanula Pyramidalis, mixed	5
Candytuft, mixed	5
" Sempervirens	10
Dianthus Chinensis, double, mixed	5
" Laciniatus, "	5
" single	5
" Plumarius, double	10
" single	10
Eschscholtzia, mixed	5
Lathyrus Latifolius, mixed	5
" white	10
Phlox Decussata, mixed	10
" large-flowering, mixed	25
Poppy Bracteatum	10
" Orientale	10
" Hybrids	25
Portulaca, single, mixed	5
" double,	10
Sweet Peas, mixed	5
" Eckford's, mixed	10
" scarlet	5
" white	5
" lavender	5
" pink	5
" 30 other varieties	5
Sweet William, mixed	5
" double, mixed	5

THE "HENDERSON" LAWN GRASS SEED.

Our Lawn Grass, sown in the Autumn, will produce a fine luxuriant Lawn EARLY the following season.

The "Henderson" Lawn Grass Seed

is the best lawn seed for our American climate. It is composed of various grasses that grow and flourish during different months of the year, so that a rich, deep green, velvety lawn is constantly maintained, rivaling the famous lawns of Old England. **The "Henderson" Lawn Seed** is the result of several years' careful experimenting, and is **unequaled.** We have made the formation of perfect and permanent lawns a study for years, and the finest lawns in this country prove our ability.

THE BEAUTIFUL LAWNS AT THE WORLD'S FAIR IN CHICAGO We produced in Six Weeks' time by sowing our

"Henderson" Lawn Grass.

Not one of the charming features at the **Exposition** was so universally admired as our "setting of emerald velvet," which, notwithstanding the long-continued drought of the summer, *retained its verdure throughout the entire season.* Besides being the only lawn grass seed granted a medal, we exhibited the separate grasses used in the mixture and received **Seven Awards** for superior quality and purity of sample. The "**Henderson**" **Lawn Grass Seed** is, relatively speaking, the cheapest seed offered, because, while the lawn grass mixtures of other seedsmen will not average over 14 lbs. to the bushel, this weighs from 20 to 21 lbs. per measured bushel.

THE QUANTITY REQUIRED.

The quantity required for making new lawns is 5 bushels per acre, or for renovating old lawns 1 to 2 bushels. For a plot 15 x 20, or 300 square feet, 1 quart is required for new, or one pint for renovating old lawns. Full printed instructions in every package and bag.

PRICES. Per qt., 25c.; per peck, $1.50; per bush., $5.00. (*If by mail, add at the rate of 5c. per quart for postage.*)

HENDERSON'S - - -
"TERRACE SOD" Lawn Grass.

A special mixture of grasses best suited for sowing on terraces, railroad embankments and side hills—grasses that produce strong, spreading roots, thus preventing heavy rains from washing them out, that will withstand drought and exposure, thrive on shallow soils, and at the same time produce a rich velvety green turf throughout the season.

Price, 35c. per qt., $2.00 per pk., $7.00 per bushel. (*If by mail, add at the rate of 5c. per quart for postage.*)

"SUNNY SOUTH" Lawn Grass.

Specially prepared for the South or for very dry sections, and contains the best grasses adapted for hot, dry situations. Price, 35c. per qt., $2.00 per pk., $7.00 per bushel. (*If by mail, add at the rate of 5c. per quart for postage.*)

HENDERSON'S "SHADY NOOK" - - - - - Lawn Grass. - - - - -

On nearly all lawns there are unsightly bare spots under the shade of trees, which have baffled all efforts to get into grass, but with this mixture there need be no further difficulty.

Price, 35c. qt., $2.00 pk., $7.00 bushel. (*If by mail, add at the rate of 5c. per quart for postage.*)

$7.00 per bushel.

HENDERSON'S - - -
"GOLF LINK" Grass Seed.

The bracing game of Golf has now become so deservedly popular here that many inquiries have reached us lately for a mixture of grasses that will produce a sward equal to those of the "GOLF LINKS" of Scotland.

While the outlying portion of the "Links" can be seeded with less expensive grasses, the "putting greens" for about twenty yards around the "holes" must be sown with the very finest grasses.

For Putting Greens we can supply a special mixture of grasses. Price, 30c. per qt., $1.75 per pk., $6.00 per bushel. (*If by mail, add at the rate of 5c. per quart for postage.*)

For the Outlying Grounds or Links proper we can supply a suitable mixture of grasses. Price, $2.50 per bushel; 20 bushels and upwards, $2.25 per bushel.

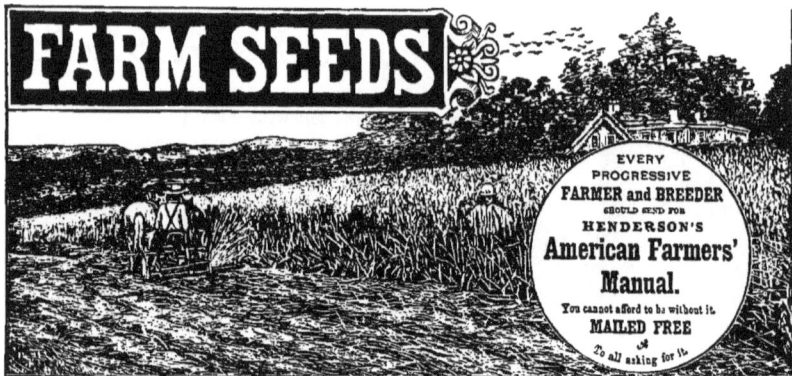

FARM SEEDS

EVERY PROGRESSIVE **FARMER and BREEDER** SHOULD SEND FOR **HENDERSON'S American Farmers' Manual.** You cannot afford to be without it. **MAILED FREE** To all asking for it.

HENDERSON'S SPECIAL GRASS MIXTURES
FOR HAY AND PERMANENT PASTURE.

. . WILL LAST TWENTY YEARS WITHOUT RENEWAL. . .

WHAT OUR CUSTOMERS SAY:

IN MAINE.
I did not succeed in getting four tons or more per acre, but I did get more than from anything else, and it is splendid hay.—G. M. HOLMES.

IN NEW JERSEY.
The Special Grass Mixtures have been most successful, in spite of an unusually trying season. I cut a crop of hay more than twice as great per acre as the timothy, which I sowed alongside at the same time.—R. V. LINDABURY.

IN IOWA.
Your Grass Mixture has done exceedingly well, especially in this year of drouth. It yielded about twice as much as timothy, and cattle seemed to do better on it.—W. WATSON.

IN NEW YORK.
Your Permanent Mixture is the finest piece of grass anywhere about here. An old farmer told me last week it would cut 3½ tons to the acre sure.—J. M. RICHARDS.

IN VERMONT.
Your Grass Seeds are easily the best of any in the market.—F. C. KIMBALL.

IN PENNSYLVANIA.
The Permanent Pasture Grass was a perfect success. We never had such a good yield of grass. We mowed our meadows twice.
—WM. SIMPSON & SONS.

The green appearance of the field attracted universal attention. It has far surpassed clover and timothy in the amount it yields.
—J. B. CUMMINGS.

IN VIRGINIA.
My manager is enthusiastic over your mixture for Hay and Permanent Pasture. He claims this year he cut 3 tons per acre the first cutting, 1½ tons the second, and will cut 2 tons the third. The field is a grand sight; people come for miles around to see it.—EDWARD E. BARNEY.

IN INDIANA.
Your Grass Seed produced the largest crop of hay I ever saw, and has yielded a most excellent pasture ever since, notwithstanding the extraordinary drouth in this vicinity.—K. S. TAYLOR.

IN OHIO.
The Special Grass Mixture has given us the very best of satisfaction; it produced the finest piece of grass I ever saw and has given us a large amount of hay and pasture per acre.—W. J. HAYES.

Consisting of the following varieties: Orchard Grass, Meadow Foxtail, Sheep's Fescue, Rhode Island or Creeping Bent, Hard Fescue, Sweet Scented Vernal (True Perennial), Meadow Fescue, English Rye Grass, Italian Rye Grass, Red Top, etc., as recommended in our book **"How the Farm Pays,"** blended in proportions which, we have found from actual use, give the most satisfactory results.

On ordinary fertile soil 3 bushels of this mixture is sufficient to seed an acre, but where the land is poor a larger quantity will be necessary. Taking one soil with another a fair average would be **3 bushels to the acre.**

For Hay and Permanent Pasture for Light soils, .		
" " " " Medium soils, .		$2.50 per bushel of 14 lbs.
" " " " Heavy soils, .		20-bush. lots, $2.45 per bush.
" Orchards and Shady Places, ...		50 " 2.40 "
" Hay only, ...		100 " 2.35 "
" Pasture only, ...		
" Renovating Old Pastures, ...		

To these mixtures, intended for either Mowing Lands or Pasture (but which on account of their greater weight should be sown separately), are to be added 10 lbs. of Mixed Clovers, comprising White, Mammoth Perennial or Cow Grass, Alsike, Trefoil, etc., but these should only be sown in the spring, as they are rather tender in this latitude if sown in the fall.

OPINIONS OF THE PRESS.

Country Gentleman says: "Below the taller grasses was a thick mat of finer kinds, and the close, rich turf hid every particle of soil."

American Agriculturist says: "Such mixtures are far superior to Timothy, or 'Timothy and Clover,' or any one grass, costing but a little more, lasting much longer, and giving frequently more than double the yield."

Farm and Home says: "The enormous yield of nearly four and one-half tons of good hay per acre should convince any one that more hay and better pasture can be grown with mixtures than with Timothy and Clover alone, as under the very same conditions the latter yielded less than a ton and a half per acre."

How the Farm Pays says: "Far in advance not only of Timothy but of any other Grass we have thus far in cultivation."

At our Farm one of these Mixtures yielded (first cutting, 5,888 lbs.; second cutting, 4,320 lbs. per acre) a total of 10,208 LBS. CURED HAY PER ACRE, while Timothy growing alongside under same conditions yielded only one cutting of 2,400 lbs. per acre.

FALL SOWING IS THE MOST SUCCESSFUL.

HENDERSON'S "RECLEANED" GRASS SEEDS.

FALL SOWING IS THE MOST SUCCESSFUL.

Prices for Grass Seed are subject to the fluctuations of the market. Those herein named are the prices ruling at this date (August), but we cannot be bound by them for any length of time. Write for prices.

Awnless Brome Grass. (*Bromus Inermis.*) A new forage plant of the utmost importance, particularly in dry and southern sections ; yields enormously. 20c. lb., $16.00 100 lbs.

Bermuda. (*Cynodon Dactylon.*) Valuable for Southern States; withstands heat and drought. $1.25 lb., 10 lbs. $11.00.

Creeping Bent Grass. (*Agrostis Stolonifera.*) Excellent for lawns; succeeds well in moist situations. About 2 bushels to the acre. (20 lbs. to bush.) 22c. lb., $3.50 bush., $18.00 100 lbs.

Crested Dog's Tail. (*Cynosurus Cristatus.*) Should enter in moderate quantity in permanent pasture mixtures and lawns. (About 21 lbs. to bush.) 60c. lb., $12.00 bush., $55.00 100 lbs.

English Rye Grass. (*Lolium Perenne.*) A valuable grass. 2½ to 3 bush. to the acre. (24 lbs. to bush.) $2.25 bush., $8.00 100 lbs.

Fine Leaved Sheep's Fescue. (*Festuca Ovina Tenuifolia.*) (About 14 lbs. to the bush.) 35c. lb., $4.25 bush., $28.00 100 lbs.

Fowl Meadow Grass. (*Poa Serotina.*) Valuable on low moist lands and meadows. Uncleaned seed. 30c. lb., $3.50 bush. of 12 lbs.

Hard Fescue. (*Festuca Duriuscula.*) Dwarf, hardy grass, of great value for dry situations; indicates superior quality in hay. (12 lbs. to bush.) 28c. lb., $2.50 bush., $18.00 100 lbs.

Hungarian Grass. (*Panicum Germanicum.*) Is a valuable annual forage plant. 1 bush. to the acre. (48 lbs. to the bush.) 10c. lb., $1.00 bush., $3.10 100 lbs.

Italian Rye Grass. (*Lolium Italicum.*) Unequaled for producing an abundance of early spring feed, giving quick and successive growths throughout the season. (18 lbs. to the bush.) 12c. lb., $1.90 bush., $10.00 100 lbs.

Johnson Grass. (*Sorghum Halapense.*) Of greatest importance for the South. Very tender and nutritious. 20c. lb., $3.50 bush. of 25 lbs., $12.00 100 lbs.

Kentucky Blue Grass. (*Poa Pratensis.*) Very valuable for a variety of soils from moist to dry, furnishes delicious and luxuriant pasturage, and makes excellent hay. (14 lbs. to bush.) 14c. lb., $1.50 bush., $10.00 100 lbs. Fancy and Double extra clean, 18c. lb., $1.75 bush., $12.00 100 lbs.

MEADOW FESCUE. (*Festuca Pratensis.*) Of great value for permanent pasture and hay; robust grower and nutritious. (22 lbs. to the bush.) 15c. lb., $2.75 bush., $12.00 100 lbs.

MEADOW FOXTAIL. (*Alopecurus Pratensis.*) Resembles Timothy, but of much earlier and rapid growth; particularly valuable for permanent pastures and hay. (7 lbs. to bush.) (*See cut.*) 40c. lb., $2.50 bush., $32.00 100 lbs.

ORCHARD GRASS. (*Dactylis Glomerata.*) One of the most valuable of all grasses, for either grazing or mowing, of early, rapid and luxuriant growth. (*See cut.*) (14 lbs. to bush.) $2.50 bush., $18.00 100 lbs.

Red or Creeping Fescue. (*Festuca Rubra.*) Suitable for sandy seacoasts and dry soils. (About 14 lbs. to the bush.) 28c. lb., $2.75 bush., $18.00 100 lbs.

Red Top Grass. (*Agrostis Vulgaris.*) (14 lbs. to bush.) $1.00 bush., $7.00 100 lbs. Recleaned seed 32 lbs. to bush.), 28c. lb., $8.00 bush., $25.00 100 lbs.

Rhode Island Bent Grass. (*Agrostis Canina.*) A very fine variety for lawns. About 3 bush. to the acre. (Bush. of 14 lbs.) 25c. lb., $2.75 bush., $18.00 100 lbs.

Rough Stalked Meadow Grass. (*Poa Trivialis.*) Valuable for pastures and meadows, particularly on damp soils. 1½ bush. to the acre. (About 14 lbs. to the bush.) 45c. lb.

Sheep's Fescue. (*Festuca Ovina.*) Excellent for uplands and dry pastures, of close, dense and nutritive growth, relished by sheep. (12 lbs. to bush.) 25c. lb., $2.50 bush., $18.00 100 lbs.

Sweet Vernal Grass, True Perennial. (*Anthoxanthum Odoratum.*) Very aromatic, giving hay a fine flavor. (10 lbs. to the bush.) $1.00 lb., $9.00 bush.

Tall Meadow Fescue. (*Festuca Elatior.*) Very early, nutritive and productive. Valuable on wet or clay soils. (About 14 lbs. to the bush.) 35c. lb., $4.50 bush.

Tall Meadow Oat Grass. (*Avena Elatior.*) Recommended for soiling, being rapid and luxuriant in its growth. 5 to 6 bush. per acre. (Bush. of 10 lbs.) 25c. lb., $2.25 bush., $20.00 100 lbs.

Timothy. (*Phleum Pratense.*) We offer a particularly "choice" sample. ½ bush. per acre. (45 lbs. to bush.) $2.25 bush., $5.00 100 lbs. Price variable.

Various Leaved Fescue. (*Festuca Heterophylla.*) 25c. lb., $3.00 bush. of 14 lbs.

Wood Meadow Grass. (*Poa Nemoralis.*) Of early growth, and thriving well under trees. 2 bush. to the acre. (About 14 lbs. to the bush.) 40c. lb., $5.00 bush.

Yellow Oat Grass "True." (*Avena Flavescens.*) Good for dry pastures and meadows. (About 7 lbs. to the bush.) $1.25 lb., $8.50 bush.

MEADOW FOXTAIL.

ORCHARD GRASS.

Send for our "Farmers' Manual." It illustrates all the best Grasses, gives full descriptions and much information upon FARM SEEDS of inestimable value to the farmer.

WINTER WHEATS.

Prices are subject to the fluctuations of the market. The prices herein named are those ruling at this date (August), for the new crop 1897. Delivery f. o. b., New York. Special quotations to large buyers.

RURAL NEW YORKER, No. 57.

Has heavily bearded heads which are beautifully symmetrical, being pointed at the tip, broad in the middle and tapering towards the stem. The straw is unusually tall and strong and stools freely, frequently having 35 to 40 stalks from a single grain. The heads are compact, averaging three kernels to a spikelet or "breast," and ten breasts to a side. The kernels are of medium size and of an attractive color, between the so-called "red" and amber. Possessing the requisite degree of hardness for the production of the finest grade of flour, it will be much sought after by millers. The chaff is clear white, with a trace of velvet sufficient to make it difficult for the green fly to attack it, and the heads do not mildew as the full velvet chaff varieties are liable to do. (*See cut.*) $1.00 peck ; $2.75 bush ; 10-bushel lots, $2.50 bush.

RURAL NEW YORKER, No. 6.

This beardless variety is a hybrid between Rye and Armstrong Wheat, though apparently all traces of Rye have disappeared and it now appears a handsome, beardless wheat. It succeeds and produces heavy crops on poor, thin land, where wheat could not be successfully or profitably grown, and it also has extreme hardiness to recommend it.

When first raised, some years ago, the top of the culms was downy with Rye culms. This characteristic could not be fixed, so that for this variety the culms having no down were alone selected. The gold-colored straw is very thick and strong, easily supporting the heavy grain without breaking. The large amber kernels are placed four to a breast, eight breasts to a side, with long symmetrical heads having a brown chaff. (*See cut.*) $1.00 peck ; $2.75 bushel ; 10-bushel lots, $2.50 bushel.

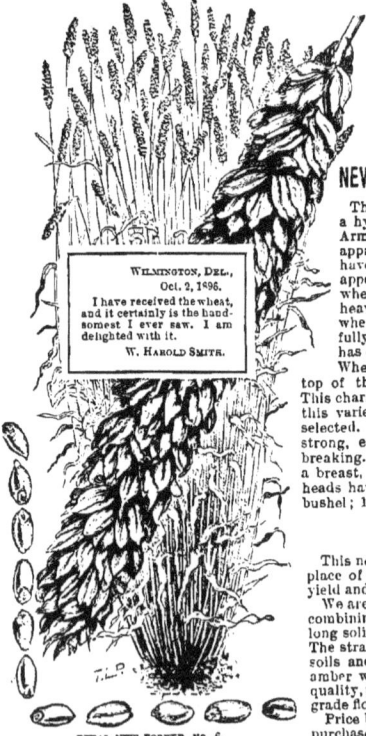

RURAL NEW YORKER, NO. 57.

RURAL NEW YORKER, NO. 6.

JONES' LONGBERRY, No. I.

This new variety, offered this year for the first time, will quickly take the place of the popular Longberries now no longer profitable, through light yield and weakened vitality.

We are confident it will prove the most profitable Longberry ever known, combining as it does strongest possible growth, strong gold-colored straw, long solid filled head, beautiful grain, and is a champion in productiveness. The straw is of such strength that it is not liable to lodge, even on strong soils and river bottoms. From the fact that it is a blending of red and amber wheat in one berry, millers will quickly recognize its high milling quality, possessing as it does the requisite hardness for the production of fine-grade flour. A trial of this grand sort will convince any farmer of its value.

Price by mail, postpaid, 1 lb., 60c.; 3 lbs., $1.50; by freight or express at purchaser's expense, $2.50 peck ; $6.00 bushel.

WINTER WHEAT—Continued.

Pride of Genesee (Bearded).—One of the most productive varieties, having a long, well-filled head, and the fact that it will give a reasonably good crop on land so poor that common sorts would be a failure, cannot fail to make it a popular sort, as the head does not decrease in proportion to the straw, being large and well filled on a very short, light growth of straw. If sown on strong wheat land, it will require less seed per acre than any other variety, 50 lbs. being ample if the field is fitted as it should be and is sown early in September. 75c. peck; $2.50 bushel; 10-bushel lots, $2.25 bushel.

Diamond Grit, or Winter Saskatchewan (Bearded).—A worthy rival at last to the Hard Spring of the Northwest, being superior to every known winter-wheat for milling, and will be the means of stimulating farmers to a more general cultivation of the winter wheat crop. With this grand seedling in general cultivation the winter wheat sections can compete with the finest grade of flour known, with the advantage of making more flour to the bushel than any other wheat as yet grown. This, with its wonderful productiveness, strong, wiry straw and sturdy growth, with extreme hardiness, cannot fail to make it a leader wherever given a trial. It is a strong grower but moderate stooler, requiring a peck more seed to an acre than most other sorts unless the land is very strong and in a fine state of cultivation. Straw is of medium height, thick-walled and wiry, of a light yellow color. Heads of medium length and carried nearly erect. $1.00 peck; $3.50 bushel; 10-bushel lots, $3.00 bushel.

Oatka Chief (Bearded).—A very strong grower even on light soils. Straw of medium height, sturdy and strong, but free from that harsh, wiry nature so common to the sturdy growers, and cannot fail to be appreciated for feeding. Beards light and short. Chaff white and very soft. Grain medium long, of light amber shade and of fine milling qualities. It is one of the most handsome wheats in the field and cannot fail to attract attention, both in field and granary. $1.00 peck; $3.50 bushel; 10 bushel lots, $3.25 bushel.

Bearded Winter Fife.—A grand new bearded Wheat, which has all the splendid milling qualities of the celebrated Winter Fife, and is even hardier and more flinty in the grain. It is one of the earliest wheats, ripening along with the Early Red Clawson, and is a very strong, healthy grower, stooling rapidly in the fall. It starts early in spring, and is among the first to head. Straw is strong and above the medium height. The heads are long and wide, with white velvety chaff. The grain is large, medium long and plump, and of a clear light amber shade. The bran is exceptionally thin. hence it will make more flour than almost any other sort grown. 75c. peck; $2.50 bushel; 10-bushel lots, $2.25 bushel.

WHEAT BY MAIL, POSTPAID.

For the benefit of our customers living at a distance from Railroads and Express Offices who would like to try our new wheats, we offer them in small lots:

		1 lb.	3 lbs.	
Postpaid	Diamond Grit, or Winter Saskatchewan	$0.50	$1.25	*Postpaid*
	Pride of Genesee	.40	1.00	
	Bearded Winter Fife	.40	1.00	
by	Jones' Longberry, No. 1	.60	1.50	*by*
	Oatka Chief	.50	1.25	
mail.	Rural New-Yorker, No. 6	.30	.75	*mail.*
	" " " No. 57	.30	.75	

OATS.

Winter.—We offer a remarkably hardy stock of Winter or Turf Oats which we have grown for six years past, being of about the same hardiness as Scarlet Clover. In favorable seasons it will winter as far north as New York, and is invariably hardy, New Jersey southwards. They produce a much heavier and longer straw than Spring Oats; stool thickly, are entirely rust-proof and never lodge. All farmers south of New York should try an acre or more. $1.25 bushel; 10 bushels and upwards, $1.10 bushel.

RYE.

Winter.—The variety most commonly cultivated. $1.10 bushel; 10-bushel lots, $1.00 bushel.

Excelsior Winter.—A new variety from Vermont, that has never failed to yield at the rate of 40 to 50 bushels per acre. With the originator, a four-acre field yielded 52 bushels to the acre. $1.50 bushel; 10-bushel lots, $1.40 bushel.

Thousandfold.—Said to be the most productive Rye in cultivation. $1.50 bushel; 10-bushel lots, $1.40 bushel.

Giant Winter.—Unquestionably the heaviest cropping Rye in existence, having in fair tests outyielded all other varieties both in straw and grain. The heads average six to eight inches in length and are filled from end to end with large, plump heavy grains. The straw is giant in length and strength and of extraordinary stiffness, resisting severe wind and rain storms to a remarkable degree without lodging. (*See cut.*) $2.00 per bushel; 10-bushel lots, $1.75 bushel.

SPECIAL QUOTATIONS TO LARGE BUYERS.

GIANT WINTER RYE.

SAND, OR WINTER VETCH.

SAND, OR WINTER VETCH

(Vicia Villosa.)

Though it succeeds and produces good crops on poor, sandy soils, it is much more vigorous on good land and grows to a height of 4 to 5 feet. It is perfectly hardy throughout the United States, remaining green all winter, and should be sown during August and September, mixed with Rye, which serves as a support for the plants.

It is the earliest crop for cutting or plowing under in spring, being nearly a month earlier than Scarlet Clover and much hardier.

It is exceedingly nutritious and may be fed with safety to all kinds of stock.

It is of such early and rapid growth that a full crop can be taken off in time for planting spring crops.

Sow one-half bushel to a bushel per acre with one-half bushel of Rye or Wheat. 15c. lb., $4.50 bushel of 60 lbs., $7.50 100 lbs.

TRUE DWARF ESSEX RAPE.

In the United States we have millions of acres of good land that annually lie idle or run to weeds the latter part of the season, after the grain, potato and hay crops have been harvested, a large portion of which might be made to produce one of the finest feeds imaginable and in the greatest abundance, at a time when cattle and sheep are roaming through bare pastures in search of a scanty living. Under favorable conditions it is ready for pasturing sheep or cattle within six weeks from time of sowing, and on an average one acre will carry twelve to fifteen sheep six weeks to two months. When on the rape they should at all times have access to salt; but water is not necessary. There are several varieties of **Rape**, but care should be taken to procure the **Dwarf Essex**, which does not seed the same season as sown. In the Northern States it should be sown from May to August for fall pasturing, but as it thrives best in cool weather, it should not be sown in the Southern States until September or October for winter pasture. In the latitude of New York, July or August is the best time to sow. Its fattening properties are probably twice as good as those of clover, and for sheep the feeding value of **Rape** excels all other plants we know of. At the Michigan Experiment Station, 128 lambs were pastured for eight weeks on 15 acres of Rape sown in July, and showed a gain of 2,890 lbs., or at the rate of 3 lbs. per lamb each week. To secure the best results the Rape should be sown in drills. Sow 6 lbs. per acre broadcast, 2 to 3 lbs. per acre in drills. 12c. lb., $9.00 per 100 lbs.

CRIMSON OR SCARLET CLOVER

(Trifolium incarnatum.)

The Most Valuable Plant for Restoring the Fertility of Worn-out Soils.

The value of Scarlet Clover is now so thoroughly established that we have no hesitation in recommending that all lands from which crops have been harvested during the summer and fall should be sown with Scarlet Clover for plowing under the following spring. Authorities who have made a careful estimate state that plowing under a good crop of Scarlet Clover is equivalent to 20 tons of stable manure per acre, and even if the Clover be harvested or pastured, the benefits derived from the wonderful nitrogenous root formation will alone many times repay the cost of seed and labor.

If intended for feeding, it should always be cut while in the young stage and never fed to stock after the crop has ceased flowering, as serious results are apt to follow the feeding of overripe Crimson Clover.

It is the cheapest source of nitrogen, and has revolutionized the methods of farming in New Jersey, Maryland and Delaware, has restored to profitable cultivation thousands of acres of poor land, and should be extensively used throughout the entire United States for sowing among corn, tomatoes, turnips, etc., at time of last hoeing, or after potatoes, melons, cucumbers, etc., have been harvested or on grain stubble and harrowed in.

In the latitude of New York, time of sowing may extend from July 15th to Sept. 15th, and further South up to Oct. Sow 15 lbs. per acre.

Choice American-grown new crop seed thoroughly recleaned and free from weed seeds, 10c. lb., $4.20 bush., $7.00 100 lbs.

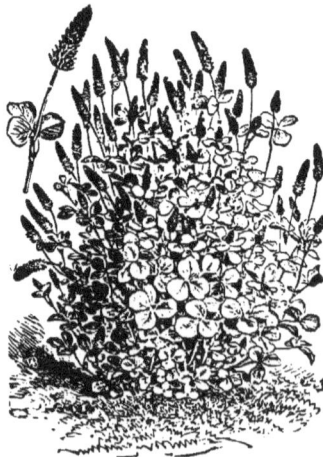

CRIMSON CLOVER.

...Portable Oil Heaters...

HEAT BY RADIATION. NO SMELL: NO DIRT: NO GAS.

Just what is wanted for small conservatories, window gardens, bathrooms, small bedrooms, etc., invaluable for protecting your plants on cold nights. We sold a large number of these heaters last season and they gave the best of satisfaction.

They are made of brass, nickel-plated, and have Russia iron cylinders, consequently there is no part of the metal that will discolor by heat. The combustion is perfect, therefore absolutely free from the offensive odor and smoke.

There is an improved Ratchet for raising wick, an improved Deflector to keep the reservoir always cool, and a Drip Cup to catch any drip down the centre draft.

The Mica Lining allows a pleasant light to shine through the open work of the cylinder.

No. 10 Portable Oil Heater. Weighs 4½ lbs., stands 23 inches high; circumference of drum, 18 inches; holds two quarts of oil; will burn 9 hours, and heat an 8 x 10 room nicely. (*See cut.*) Price, $3 50.

No. 44 Portable Oil Heater. Weighs 8 lbs., stands 32 inches high; circumference of drum, 25 inches; holds four quarts of oil; will heat a room 15 x 20 feet to a temperature of 70 degrees in the coldest weather, at a cost of about one cent per hour; has a door for lighting; will burn from 18 to 20 hours. Price, $6.00.

Revolving Adjustable Plant Stand.

No. 2. Has two tiers of brackets and holds 17 pots. Diameter, 26 inches; height, 4 feet; weight, about 25 lbs. Price, $5.00.

No. 3. (*See cut.*) Has three tiers of brackets and holds 23 pots. Diameter, 32 inches; height, 5 feet; weight, 35 lbs. Price, $6.50.

Henderson's Wire Plant Stands.

These "knock down" for shipment as shown in cut below. Consequently they reach their destination in perfect condition at small cost for transportation. Full directions accompany each, so any person can put them together. They are very strong and steady and are handsomely finished in green enamel and gold.

LARGE "HALF CIRCLE" STAND.

45 inches wide; 28 inches deep; 42 inches high, or with trellis, 72 inches high. Price, without trellis, $7.25. With trellis, $8.75.

SMALL "HALF CIRCLE" STAND.

42 inches wide; 26 inches deep; 42 inches high, or with trellis, 67 inches high. Price, without trellis, $6.25. With trellis, $7.75.

THREE SHELF SQUARE STAND.

41 inches long; 25 inches deep; 42 inches high, or with trellis, 75 inches high. Price, without trellis, $7.25. With trellis, $8.75.

TWO SHELF SQUARE STAND.

34 inches long; 17 inches deep; 32 inches high, or with trellis, 60 inches high. Price, without trellis, $4.50. With trellis, $6.00.

Small Shelf Square Stand.

28 inches long; 10 inches deep; 22 inches high. Price, $3.50.

NO. 44 PORTABLE OIL HEATER.

NO. 10 PORTABLE OIL HEATER.

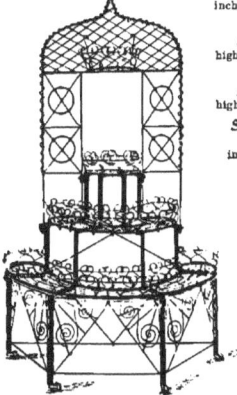

SMALL SINGLE SHELF STAND.

HALF CIRCLE STAND, WITH TRELLIS.

SQUARE STAND, WITHOUT TRELLIS.

WIRE PLANT STAND. PACKED FOR SHIPMENT.

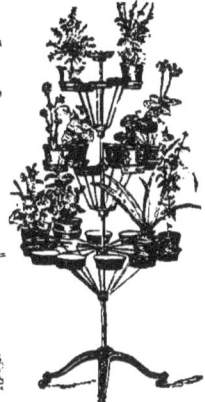

REVOLVING ADJUSTABLE STAND.

PREPARED POTTING SOIL.

One of the principal elements of success in the growing and flowering of plants and bulbs is proper soil. No matter how much care and attention is given afterwards, if the soil is not right—best results cannot be attained. It is so often difficult for amateurs to procure the right kind of potting soil that we have concluded, after many solicitations, to furnish it to our customers at practically cost to us, including packing.

This "Prepared Potting Soil" will be the same that we use in our extensive greenhouses. It takes a year or more to prepare it by first placing a layer of inverted sod, covering it with a layer of rotted manure, then another layer of sod and so on. This "mold pile" is allowed to stand about 6 months, then it is thoroughly turned and mixed, and the operation repeated several times during the last 6 months—when it is considered fit to use. As needed we mix with it peat, sand and pure raw ground bone meal—it is then friable, rich, and will cause plants to grow and bloom luxuriantly.

PRICE, 25c. per peck bag, 75c. per bushel bag, $2.00 per barrel (purchaser to pay freight).

PRUNING KNIVES.

Made of the finest steel, with stag handles.

BUDDING KNIVES.

Made of the finest quality steel, with ivory handles.

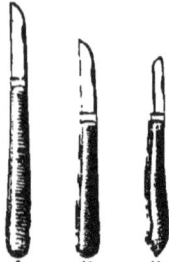

No.	13	14	15	9	10	11
Price,	$1.00	75c.	85c.	$1.00	$1.00	$1.00

RUBBER PLANT SPRINKLERS.

Indispensable article for sprinkling cut flowers, seedlings, clothes, etc. A pressure on the bulb ejects the water in a fine spray.

PLANT SPRINKLERS.

PRICES.

Straight neck, 6-oz. size, 45c.; 8-oz. 55c.; 10-oz., 65c.

Angle neck, 6-oz., 50c.; 8-oz., 65c.; 10-oz., 75c.

(Postage, 10c. each extra.)

ACRILLA'S PATENT APPLICABLE AVAILABLE

IMPROVED RUBBER PUTTY BULB.

IMPROVED RUBBER PUTTY BULBS.

An excellent article for glazing. A pressure with the hand ejects the putty prepared as per directions accompanying each, and by running it along the sash bars the work is done quickly, and makes a durable, tight and neat joint. Does not daub the glass nor stick to the hand. Price, $1.00 each; by mail, $1.10 each.

VAN REYPER'S GLAZING POINTS.

The finest glazing points on the market; the glass cannot slip; they are quickly and easily put in with special pincers; no "rights and lefts" to bother with. Price per box of 1,000, 65c., or by mail, 80c.

GLAZING POINTS.   Pincers, price pair, 50c., or by mail, 60c.

EUREKA FUMIGATORS.

GLAZING POINTS.

For fumigating greenhouses with dampened tobacco stems; made of galvanized sheet iron, a damper regulates the draft; no danger of fire; no ashes or litter. (See cut.) No. 1, 12 in. high, $1.50; No. 2, 16 in. high, $1.75; No. 3, 20 in. high, $2.00; No. 4, 24 in. high, $2.75.

GLOVES.

Gardeners' Gloves, made of heavy sheepskin, saves the hands while working among thorny plants. Per pair, $1.00, or by mail, $1.10.

GARDENERS' GLOVES.

Rubber Gardening Gloves, in black, white, EUREKA FUMIGATOR tan or lavender, with half long gauntlets. Per pair, $1.50, or mailed, $1.60. Ladies' sizes, Nos. 6, 7, 8, 9. Men's sizes, Nos. 10, 11, 12.

FLORENCE VAPORIZER.

A very useful article for spraying strong fluids such as solutions of Fir Tree Oil, Coles' Insect Destroyer, etc., on house and garden plants. Price, 60c. each, or by mail, 75c.

JUMBO POWDER GUN.

For applying insect powders on plants in the house or small gardens; it will hold about 4 oz. of powder, which is ejected and distributed by a pressure with the thumb. The "Jumbo" we consider the best small powder distributer on the market. (See cut.) Price, 20c. each, or by mail, 25c. each.

INSECT POWDER GUN JUMBO

SCISSORS AND SHEARS.

FLOWER GATHERING.	GRAPE THINNING.	PRUNING, SOLID STEEL.	LOPPING.	HEDGE.
75c. Mailed, 80c.	65c. Mailed, 70c.	7½ in. 80c.; 8½ in. 90c.; 9½ in. $1.00.	$2.00.	8 in. $1.25; 9 in. $1.50; 10 in $1.75 Notch extra, 25c.

BRASS SYRINGES.

These Syringes are applicable for all horticultural purposes in the conservatory and garden. They are fitted with caps or roses for ejecting water in one stream, or dispersing it in a fine or coarse spray as required. Specially adapted for applying fluid insecticides.

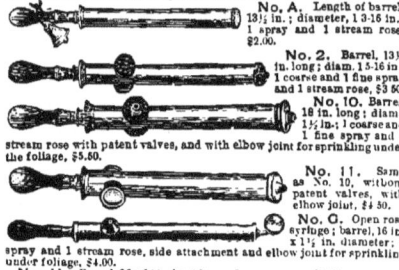

No. A. Length of barrel, 13½ in.; diameter, 1 3-16 in.; 1 spray and 1 stream rose, $2.00.

No. 2. Barrel, 13½ in. long; diam. 1 3-16 in.; 1 coarse and 1 fine spray and 1 stream rose, $3.50.

No. 10. Barrel, 18 in. long; diam., 1½ in.; 1 coarse and 1 fine spray and 1 stream rose with patent valves, and with elbow joint for sprinkling under the foliage, $5.50.

No. 11. Same as No. 10, without patent valves, with elbow joint, $4.50.

No. C. Open rose syringe; barrel, 16 in. x 1½ in. diameter; 1 spray and 1 stream rose, side attachment and elbow joint for sprinkling under foliage, $4.00.

NO. H. Barrel, 18 x 1¼, sheet brass, 1 stream rose, $2.25.

MASTICA FOR GLAZING

Greenhouses, Sashes, etc., new and old. It is Elastic, Adhesive and easily applied; it is not affected by dampness, heat or cold.

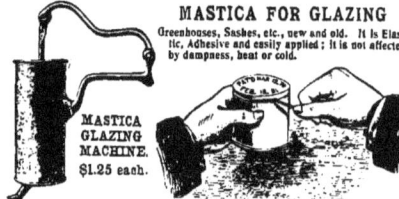

MASTICA GLAZING MACHINE. $1.25 each.

Every florist has experienced difficulty in obtaining putty (whether ordinary or white lead), for glazing, that is satisfactory for any length of time. The fact is, putty is not adapted for greenhouse work.

After much study the inventor of "Mastica" decided that the composition must be of different material, and the qualities must be elastic and tenacious to admit of expansion and contraction without cracking. "Mastica," when applied, in a few hours forms a skin or film on the entire mass, hermetically sealing the substance and preventing the evaporation of the liquids, and remains in a soft, pliable and elastic condition for years. "Mastica" is of great value in going over old houses with a putty bulb or machine on the outside of sash as it makes it perfectly tight and saves the expense of relaying the glass.

Prices of Mastica. (Soft for machine application), 50c. per quart; 75c. per ½ gallon; $1.25 per gallon.

RAPHIA.

The best and cheapest tying material for plants, vegetables, grafts, etc.

In braided plats.

PRICE.

20c per lb.; 10 lb. for $1.50.

(Postage, extra, 16c. per lb.)

RAPHIA FOR TYING

BURLAP MATS
FOR PROTECTING COLD FRAMES AND
HOTBEDS.

BURLAP MATS,
For Protecting Cold Frames and Hotbeds.

These are 6x6 feet square, made of strong burlap bagging, warmly lined with waste wool and cotton, which is quilted in to hold position. They are excellent substitutes for straw mats, being, if anything, warmer than straw, more easily handled, less bulky, and they do not harbor mice or other vermin. We were afraid that they would hold moisture, and either rot or mildew, but our trial for two winters proves them to be far more durable than straw mats.
Price, $1.00 each, $11.00 per dozen.

STRAW HOT-BED MATS.

Made of rye straw and tarred cord. They are invaluable for throwing over cold frames, hotbeds, etc., during the coldest weather. Price, size 3 x 6 feet, $1.25 each, $14.00 per doz.; size 6 x 6 feet, $2.00 each, $22.00 per doz.

PATENT PROTECTING CLOTH.

Specially prepared to prevent mildewing and rotting; valuable for protecting plants from frosts, covering hotbeds and frames in spring, in lieu of glass, for Chrysanthemum houses, for covering tender bedding plants at night when there is danger of frost, thereby lengthening the display, etc., at one-tenth the cost of glass. *Samples and circular mailed on application.*

TEMPORARY CHRYSANTHEMUM
HOUSE OF PROTECTING CLOTH.

PRICES:

Heavy Grade, per yd., 12c.; per piece of 40 yds., $4.50; weighs about 46 lbs. per 100 yds.
Medium Grade, *best for general purposes*, per yd., 10c.; per piece of 50 yds., $4·25; weighs about 2½ lbs. per 100 yds.
Light Grade, *mostly used in South for tobacco plants*, per yd., 6c.; per piece of 60 yds., $2.50; weighs about 1½ lbs. per 100 yds.

HOTBED SASH.

Unglazed, best quality, size 3 x 6 feet, using 6 x 8 inch glass. Price, 80c. each; per doz., $9.00. In shooks (not put together), 70c. each, $8.00 per doz.

DEAN'S AUTOMATIC MOLE TRAP.

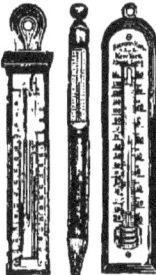

The Best Trap Made.

It is easier set than any other trap. Simply pull the plunger and it sets itself. It will catch moles when quite deep in the ground, and there being no pin or other obstruction projecting into the run, there is nothing to frighten the mole as it passes, and in doing so raises the ground enough to spring the trigger. The points of the pins being in the ground, it cannot injure domestic animals.
Price, $1.00 each.

DEAN'S
AUTOMATIC MOLE
TRAP.

Thermometers.
Cannot be mailed.

SIEXE'S. HOTBED. PARLOR.
HOTBED.
THERMOMETER.

Siexe's Self-Registering. Registers both heat and cold; enameled case, 8-inch, $2.75.
Greenhouse Distance. Plain, readable, porcelain scale, ranging from 20 to 100 deg., $1.00.
Parlor, Boxwood case, metal scale, 12-inch, 50c.
Ebony, Porcelain scale, 8-inch, $1.00; 10-inch, $1.50; 12 inch, $2.00.
Plate Glass, With brass supporters, 8-inch, $1.50; 10-inch, $1.75.
Ordinary Japanned Tin Cases. (*Not guaranteed correct.*) 7-inch, 15c.; 8-inch, 20c.; 10-inch, 25c.; 12-inch, 30c.
Hotbed and Mushroom-bed. Brass-mounted and tipped, made especially for plunging, $2.00.

THE "PLUMLEY" FRUIT PICKER.

A very simple device, without springs or machinery to get out of order or injure the fruit, which is pulled off by three curved iron fingers, and it rolls easily down a cloth tube, the bottom of which can be held in a basket or barrel. By holding the cloth tube with one hand the fall of each fruit is checked, and can let out as easily as the operator desires, preventing bruising. As the picker does not have to be lowered with each fruit, as with some pickers, the result is ten times as much can be picked, and with greater ease, by one person. The length of the tube is 11½ feet, enabling a man of ordinary height to reach fruit 16 feet from the ground, and as much lower as desired by having poles of different lengths. We do not furnish poles.

THE
"PLUMLEY"
FRUIT PICKER.

Price (*without pole*), $1.00 each, or prepaid by mail, $1.15.

WIRE FRUIT PICKER.

A very excellent device; can be attached to poles of any length.
Price, without pole, 25c.; by mail, 35c.

"DOUBLE EDGE" PRUNING SAW.

Cuts by direct action both ways. Holes are provided so it can be attached to a pole. The "lightning" teeth are used for the larger limbs, the fine teeth for the smaller limbs. Price, 16-inch, 60c.; 18-inch, 70c.; 20-inch, 80c.

COMBINED SAW AND CHISEL.

The saw is twelve inches long. The chisel is three inches wide, made thin. A handle of convenient length can be inserted. Price, 90c.

WATERS' IMPROVED TREE PRUNER.

WATERS
IMPROVED
TREE PRUNER.

The hook encircles the limb and supports the blade on both sides, which allows the blade to be made very thin, thereby reducing the resistance of the wood, and making an easier and smoother cut than any other device. The small space required for working the knife allows it to be used among close, dense branches.
Price, length of pole, 4 ft., 75c.; 6 ft., 85c.; 8 ft., $1.00; 10 ft., $1.15; 12 ft., $1.25. Extra knives, each, 20c.

TREE TUBS, CEDAR.

Green outside and brown inside. (*Painted red, 10 per cent. extra.*)

No.	Outside Diameter.		Length of Stave.		Price.
0........	27 inches	....	24 inches	......	$7.00
1........	25 "	....	22 "	......	6.00
2........	21 "	....	20 "	......	5.00
3........	21 "	....	18 "	......	4.00
5........	18 "	....	14 "	......	3.50
6........	16 "	....	14 "	......	3.00
6........	14 "	....	12 "	......	2.50
7........	13 "	....	11 "	......	2.25
8........	12 "	....	10 "	......	2.00

Window Brackets for Plants.

Highly finished and supplied complete with screws.

	Each.
1 pot	$0.25
2 "	.50
3 "	.75
4 "	1.00

HANGING BASKET AND BIRD CAGE HOOK, 15c. each.

4-POT BRACKET.

HYACINTH GLASSES.

Hyacinths grown in glasses of water are charming ornaments for parlor windows, and the ease and success in flowering bulbs in this way add greatly to their popularity. Polyanthus Narcissus, Amaryllis Formosissima and a few other bulbs can be grown in the same manner.

VERNAL

BOHEMIAN FANCY.

TYE'S SHAPE.

TALL.

JAPANESE.

PATENT.

Vernal Hyacinth Vase. Made of terra cotta. Immerse the vase for a day in water, then sprinkle it with Timothy seed; within a few days' time will cover the vase with a bright green growth; of course, always keep the vase filled with water. Hyacinths can be grown in these just the same as in the regular Hyacinth glasses. Price, 30c. each, $3.00 per doz. 1 oz. Timothy seed, 10c.

Tye's Shape Hyacinth Glass. Various colors. 15c. each, $1.50 per doz.

Tall Hyacinth Glass. Various colors. 15c. each, $1.50 per doz.

Bohemian Hyacinth Glass. Various colors. 30c. each, $3.00 per doz.; in engraved or gilt designs, 50c. each, $5.00 per doz.

Patent Hyacinth Glass. This is in two parts; the inner contains the bulb and the roots, and can be removed without injuring them, to refill with water. 25c. each, $2.50 per doz.

Japanese Hyacinth Glass. White porcelain decorated in colors and gold, under glaze in Japanese designs. 40c. each, $4.00 per doz.

CHINESE Sacred Lily BOWLS.

As used by the Celestials for growing their famous "Joss Flower" in. The bulbs should be surrounded with pebbles to keep them from toppling over when in flower, and then water enough put in to cover about one-half of the bulb; place them in a dark, cool closet for a couple of weeks to become rooted, and then bring into the light.

Tokio Chinese Lily Bowls.
China, handsomely decorated in curious Japanese designs.

7-inch, for 1 bulb $0.40
8   "   2   "   .... .60
9   "   3 or 4 bulbs .... .80

Glass Chinese Lily Bowl.
6-inch diameter, for 1 bulb....$0.25
7½   "   "   2   " .... .30
8½   "   "   3   " .... .35

THE NEW WAY.   THE OLD WAY.

"THE ROYAL" POT LIFTER.

"THE ROYAL" POT LIFTER.

Those who have experienced trouble and had accidents in lifting potted plants out of fancy jardinières as they empty stale water, will appreciate the value of our Royal Pot Lifter. It is made of brass, is strong and will not rust.

There are various contrivances on the market for this purpose, but they are cheap wire arrangements liable to slip, rust and break, and besides, look unsightly connected with an aristocratic jardinière, while our "Royal" adds richness. We furnish them to fit pots of the following sizes: 7, 8, 9, 10, 11 and 12 inch. **Price,** 60c. each, or by mail, 65c. each.

NOVELTIES in which to grow BULBS.

Interesting and beautiful objects for the window garden may be produced by growing Crocus, Lily of the Valley and Roman Hyacinths in the ornamental pots here illustrated, which are pierced with holes, out of which the leaves and flowers grow; the bulbs are placed inside with the crowns in the holes; soil is then filled in compactly which holds the bulbs in place; soak thoroughly with water, place in a dark, cool place for a few weeks, to allow the bulbs to root thoroughly. Never allow them to suffer for water.

Roman Hyacinth Pot.
Price, $1.00 each.
(Requires 20 bulbs.)

Bee Hive Crocus Pot.
Price, 80c.
(Requires 50 Crocus.)

Hedgehog Crocus Pot.
Price, 80c.
(Requires 50 Crocus.)

Lily of the Valley Pot.
(Rough Finish.)
Price, 30c. each, $3.00 per doz.
(Requires 75 Single Crowns.)

"Tokoniba" Jardiniere.

A Japanese ware somewhat resembling very dark terra cotta; it is very hard and exquisitely shaded, dragons and birds in raised relief on dotted clouds; the effect is quiet and rich.

Prices: 8½-inch, $1.25; 9½-inch, $1.50; 10½-inch, $2.00; 12-inch, $3.00.

"Rococo" Jardiniere.

This line is exquisite, the quality of the porcelain is extra, and the decorations in various tinted colors and gold stipple and lines unsurpassed.

Prices: 6-inch, $1.50; 7-inch, $2.00; 8-inch, $2.60; 10-inch, $5.00.

Japanese Jardinieres.

(Various Shapes.)

Heavy "Seidai" porcelain, various ground colors, with flowers and decorations, under glazed enamel.

Prices: 9-inch, $1.25; 10-inch, $2.00; 12-inch, $2.75.

Surrey Jardinieres.

Tinted and ornamented with flowers in colors, gold stipple and scroll.

Prices: 7-inch, $1.00; 8-inch, $1.25; 9-inch, $1.75; 10-inch, $2.50; 11-inch, $3.50; 12-inch, $4.50.

HANGING POTS.

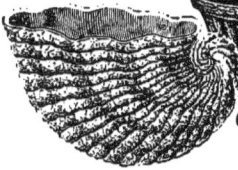

Hanging Shell.
Ivory white. 11-inch, 90c.

Hanging Round Rustic.
7-inch, 30c.; 9-inch, 40c.

Hanging Log.
7-inch, 30c.; 9-inch, 45c.

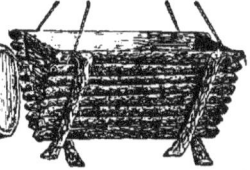

Hanging Rustic Cottage.
8-inch, 75c.; 10½-inch, $1.00.

Brass Chains, extra, each, 15c., $1.50 per dozen; by mail, 18c., $1.75 per dozen.

"Elegant" Fern Dish.

Prices include inside pan.

Fine white porcelain, with raised scroll work and legs, decorations in colors with gold stipple and border. 8-inch, $1.75 each; 9-inch, $2.00.

Minton Tile Window Box.

This beautiful box is made of glazed Minton Tiles, substantial polished oak frame, interior lined with zinc. Size, outside measurements, 33½ inches long by 9½ inches wide, by 10 inches deep; very cheap.

Price, $5.50 each. Larger size, $6.50.

ROLLING STANDS FOR HEAVY PLANTS.

A very useful waterproof saucer arrangement on casters for turning or moving heavy plants, and preventing injury to carpets from drip or dampness.

Size A, 13 inch diameter, on 3 casters, $0.75
"   B, 17 "   "   4 "   1.00
"   C, 21 "   "   6 "   1.50

"STANDARD" FLOWER POTS.

☞ No order filled for less than $2.00 worth. We pack carefully, but will make no allowance for breakage.

Breakage is not one-half as great as in other pots, the deep rim protecting them. The foot keeps the pot up from bench, so that it is impossible for hole to become clogged. The concave bottom and large hole insure perfect drainage.

STANDARD FLOWER POTS.

		Per doz.	Per 100
2	inch	$0.12	$0.75
2½	"	.15	.90
3	"	.20	1.25
4	"	.35	2.00
5	"	.45	2.75
6	"	.75	4.25
7	"	1.25	6.50
8	"	1.50	9.00
10	"	3.00	18.00
12	"	4.00	27.00

ROUND
SEED OR LILY PANS.

H'g't. Width.	Per doz.
4 x 8 inches	$1.60
5 x 10 "	1.75
6 x 12 "	2.00
7 x 14 "	2.25
8 x 16 "	2.60
9 x 18 "	3.75

SQUARE SEED PANS.

		Per doz.
6 x 6 inches		$2.00
8 x 8 "		2.50
10 x 10 "		3.00
12 x 12 "		3.50

Common Flower Pot Saucers.

		Per doz.
4 inch		$0.25
5 "		.30
6 "		.35
7 "		.40
8 "		.50
9 "		.88
10 "		1.00
11 "		1.25
12 "		1.50
14 "		1.75

WATERPROOF FLOWER POT SAUCERS.

If you have ever been annoyed with ordinary flower pot saucers absorbing moisture and dispensing it imperceptibly until the article it sat on was ruined, you will then fully appreciate the value of these new indurated fibre ware saucers; they do not absorb moisture and are not easily broken. Color, terra cotta brown.

Prices:

6 inch, 8c. each; $0.80 per dozen.
8 " 9c. " .90 "
10 " 10c. " 1.00 "
12 " 12c. " 1.20 "

REMEDIES for INSECTS & FUNGUS

Our new book, "Injurious Insects and Plant Diseases, with Remedies," gives the latest treatments and best apparatus for applying fluid and powder insecticides. Price, 25 cents.

Ant Destroyer, ½ lb. can, 25c., or prepaid, 30c.; 1 lb. can, 75c., or prepaid $0.90
Carbolic Soap, 1 lb. can, 15c., or prepaid .... .30
Fir Tree Oil, per ¼ pt., 40c.; pt., 75c.; qt., $1.40; ½ gal., $2.50; 1 gal. 4.25
Fir Tree Oil Soap, ½ lb. tin, 25c., or prepaid, 35c.; 2 lb., 75c., or prepaid, $1 05; 6 lb., $1.75; 10 lb., $3.25; 20 lb. 6.00
German Caterpillar Lime, 5 lb. can, $1.00; 10 lb. can, $1.75; 25 lb. keg, $3.75; 50 lb. keg, $6.75; 100 lb. keg 12.75
Persian Powder, High Grade, per lb., 40c., or prepaid .... .55
Kerosene Emulsion, 1 qt., 30c.; 1 gal., 50c.; 5 gals. 1.75
Paris Green, per lb., 30c., or prepaid .... .45
Hellebore, Powdered White, per lb., 20c., or prepaid .... .35
"Rose Leaf" Insecticide (Extract of Tobacco). This we find one of the most effectual insect destroyers in our greenhouses. It is to be diluted with water and can be applied in several ways, either by vaporizing through a bellows or syringing with spray nozzle, or by painting the pipes or evaporated in pans either on pipes or on oil stoves. Full descriptive pamphlet mailed on application. **Price,** pint can, 35c.; quart can, 60c.; gallon can, 1.75

Slug Shot, 5 lb. pkg., 30c.; 10 lb. pkg. $0.50
"   "   barrel of 235 lbs. in bulk, $8.50; keg of 125 lbs. 5.00
"   "   in canisters holding ½ lb. each, 25c., or prepaid ...... .35
Tobacco Soap, per lb., 35c., by mail .... .50
Tobacco Dust, per lb., 10c.; 5 lb. pkg., 35c.; 10 lb. pkg., 65c.;
if wanted prepaid, add 15c. per lb.; per bbl., $3.00; per ton 25.00
Whale Oil Soap, 1 lb. box, 15c., or prepaid, 30c.; 2 lb. box, 25c., or prepaid, 50c.; 5 lb. box, 50c.; 25 lbs. and over, 8c. per lb.
Tobacco Stems, 50 lb. bale, $1.00; 100 lb. bale 1.75
"   "   per ton of 2,000 lbs. 20.00

REMEDIES FOR FUNGUS DISEASES.

Bordeaux Mixture, ingredients ready for mixing. 1 qt. can, 50c.;
1 gal. can, $1.25; 5 gal. can 5.00
Copperdine, (Ammoniacal Solution of Carbonate of Copper, concentrated), 1 qt. can, 50c.; 1 gal., $1.50; 10 gal. carboy, at $1.35 gal.
Flour of Sulphur, per lb., 10c., or prepaid, 25c. per lb.; per 10 lbs., 65c.; 25 lbs. and over, at 5c. per lb.
Fostite, 5 lb. pkg., 60c.; 25 lbs., $2.00; 50 lbs., $3.50; 100 lbs. 6.50

FERTILIZERS

Grinding Bones for Fertilizers

Prices quoted are subject to change without notice. All quotations are per ton of 2,000 lbs.

Henderson's Superior Fertilizer for Vegetables and Flowers.
per 5 lb. pkg. $0.25
"   10   " .... .45
"   25 lb. bag 1.00
"   50   " 1.75
" 100   " 3.25
"   ton 50.00
Henderson's Lawn Enricher.
per 5 lb. pkg. .... .30
"   10   " .... .50
"   25 lb. bag 1.25
"   50   " 2.00
" 100   " 3.75
"   ton 60.00
Ground Bone, Pure, per 100 lb. bag. 2.25
"   "   "   "   200   " 4.00
"   "   "   "   ton 37.00

Pure Bone Meal, per 1 lb. pkg. $0.10
(If wanted prepaid, add 15c. per lb.)
per 5 lb. pkg. .... .30
"   10   " .... .50
"   25 lb. bag 1.00
"   50   " 1.60
" 100   " 2.75
" 200   " 4.50
"   ton 40.00
Crushed Bone, Pure, per 100 lb. bag. 2.75
"   "   "   "   200   " 5.00
"   "   "   "   ton 43.00
Nitrate of Soda, per 5 lb. pkg. .... .30
per 10 lb. pkg. .... .50
"   25 lb. bag 1.25
"   50   " 2.00
" 100   " 3.50
" 300   " 9.00

Pure Pulverized Sheep Manure.
per 2 lb. pkg., 15c., or prepaid $0.45
"   4   " .... .25
"   10   " .... .50
" 100 lb. bag 2.50
"   ¼ ton (5 bags) 10.00
"   ton (20 bags) 32.00
Blood and Bone Fertilizer.
per 100 lb. bag. 2.50
"   200   " 4.25
"   ton 39.00
Bone Superphosphate, per 100 lb. bag, 3 60
per ton 32.00
Unleashed Canada Ashes.
per bbl. of about 200 lbs. 2.25
"   ton in barrels 20.00
Asparagus Salt, per 200 lbs. 1.25
"   "   "   " 9.00

MAPES' CELEBRATED FERTILIZERS. Prices and Descriptive Pamphlet on application.

HENDERSON'S SUPERIOR FERTILIZER FOR HOUSE PLANTS.

A safe, clean and high-grade fertilizer, free from disagreeable odor, prepared especially for feeding plants grown in pots. It is a wonderful invigorator, producing luxuriant, healthy growth of rich texture and larger and more brilliant flowers of improved substance.

It contains in a highly concentrated form all of the ingredients of plant food essential to the highest development of plants and flowers. It is very soluble and is readily assimilate, so that marked improvement is usually noticed in ten days' time. It is fine and dry, clean and easy to apply, either sprinkled over the surface of the soil as a top dressing, or dissolved in water (*stirring well*). Detailed directions on each package.

For Cut Flowers, a pinch of this fertilizer in the water will keep it sweet and wholesome, and preserve the flowers from one to three days longer. If half an inch of the stems is cut off daily, which removes the callus which forms over the cut ends and allows the flower above to absorb the invigorating liquid

Price, 1 lb. package, sufficient for 25 ordinary sized plants for 1 year, 20c., or prepaid, 35c.

Henderson's New Handbook of Plants and General Horticulture

Gives a short history of the different genera, and concise instructions for their propagation and culture, and embraces the botanical name (accentuated according to the latest authorities), derivation, natural order, etc. A valuable feature of the book is the leading local or common English names, together with a comprehensive glossary of Botanical and Technical terms. Instructions are also given for the cultivation of the principal vegetables, fruits and flowers, with very full instructions on forcing Tomatoes, Grapes, Cucumbers, Mushrooms, Strawberries, etc., together with comprehensive practical directions about soils, manures, roads, lawns, draining, implements, greenhouse buildings, heating by steam and hot water, propagation by seeds and cuttings, window gardening, shrubs, trees, etc. In short, everything relating to general horticulture is given in alphabetical order to make it a complete book of reference. 526 pages, 800 illustrations. Price, $4.00, postpaid, *or given free as premium on an order. See 2d page cover.*

GARDENING FOR PLEASURE.

NEW EDITION. Tells how to grow Flowers, Vegetables and Small Fruits in the Garden and Greenhouse; also treats fully on Window and House Plants, the Lawn, Bulbs, Aquatic Plants, Modes of Heating, Small Fruits, Insects, the Grapery, Monthly Calendar of Operations. 404 pages; fully illustrated. Price, $2.00, postpaid, *or given free as premium on an order. See 2d page cover.*

GARDENING FOR PROFIT.

NEW EDITION. A new, revised and greatly enlarged edition of this popular work. This book gives in detail our 25 years' experience in *Market Gardening*, and a revised list of varieties in vegetables recommended for market culture. Written particularly for the Market Gardener, but is equally as valuable for the Private Gardener. 375 pages; fully illustrated. Price, $2.00, postpaid, *or given free as premium on an order. See 2d page cover.*

GARDEN AND FARM TOPICS.

Contains essays on some special Greenhouse, Vegetable and Bulb, Fruit and Farm Crops. 244 pages; fully illustrated. Price, $1.00, postpaid, *or given free as premium on an order. See 2d page cover.*

CULTURE OF WATER LILIES.

42 pages; illustrated. Price, postpaid, 25 cents, *or given free as premium on an order. See 2d page cover.*

PRACTICAL FLORICULTURE.

NEW EDITION. Written particularly for the Commercial Florist, but equally as valuable for the Amateur. This work teaches how flowers and plants can best be "grown for profit." It is admitted to be the leading authority on the subject. 325 pages; fully illustrated. Price, $1.50, postpaid, *or given free as premium on an order. See 2d page cover.*

HOW THE FARM PAYS.

By Messrs. Henderson and Crozier. An acknowledged authority for Farmers. Gives all of the Latest Methods of growing Grass, Grain, Root Crops, Fruits, etc.; and all about Stock, Farm Machinery, etc., etc. 400 pages; fully illustrated. Price, $2.50, postpaid, *or given free as premium on an order. See 2d page cover.*

CONDENSED VEGETABLE AND FLOWER SEED CULTURE.

An eight-page pamphlet, containing, in a condensed form, instructions for the cultivation of Garden Vegetables and Flowers from seeds. Also, full directions for making Hotbeds and Cold Frames. Price, 10c., *or given free as premium on an order. See 2d page cover.*

INJURIOUS INSECTS AND PLANT DISEASES, WITH REMEDIES.

76 pages; illustrated. Price, 25 cents, postpaid, *or given free as premium on an order. See 2d page cover.*

...HENDERSON'S BULB CULTURE...

Tells how to grow bulbs for winter flowering in the house or greenhouse, and for spring flowering in the garden. It also treats on summer flowering bulbs; in fact, gives full instructions when and how to plant, and after-management. It also tells how to "force" bulbs.

It tells how to grow bulbs in glasses, crocus pots, moss and other novel ways. It gives designs for beds of bulbs and tells what to put in them. It gives a list of common names; tells what bulbs are suitable for naturalization, for bedding, for winter flowering, for summer flowering, etc., etc., and much other valuable information. 24 pages; illustrated. Price, postpaid, 25 cents, *or given free as premium on an order. See 2d page cover.*

Special Offer: If ordered at one time, we will supply the full set of ten books, described above, carriage prepaid, for $10.00. (Separately, they would cost $13.95.) This set of books form A COMPLETE LIBRARY OF THE GARDEN, GREENHOUSE AND FARM. **For $10.00**